Electromagnetic Fields in Multilayered Structures

Theory and Applications

For a complete listing of the *Artech House Antenna Library*,
turn to the back of this book

Electromagnetic Fields in Multilayered Structures

Theory and Applications

Arun K. Bhattacharyya

Artech House
Boston • London

Library of Congress Cataloging-in-Publication Data
Bhattacharyya, Arun
Electromagnetic Fields in Multilayered Structures: Theory and Applications/Arun Bhattacharyya
Includes bibliographical references and index.
ISBN 0-89006-651-5
1. Microwave devices—Materials. 2. Dielectrics. 3. Electromagnetic fields.
4. Microwave antennas. 5. Microwave integrated circuits. I. Title.
TK7876.B52 1993 93-38271
621.381'3–dc20 CIP

A catalogue record for this book is available from the British Library

685 Canton Street
Norwood, MA 02062

International Standard Book Number: 0-89006-651-5
Library of Congress Catalog Card Number: TK7876.B52 1993

10 9 8 7 6 5 4 3 2 1

For my mother

Contents

Preface

This book is written to fulfill the need of a self-contained text that provides a technique to find the electromagnetic fields in a layered medium. Such analysis would be useful to electromagnetic engineers and scientists, especially to those who have been working in the area of microstrip antennas and microwave integrated circuits. A major portion of the material presented in this book has been published in various journals and at conference proceedings.

The book is designed for use in a graduate-level course. The reader of this book is assumed to have a working knowledge of basic electromagnetic theory and engineering mathematics at the undergraduate level. To make the book self-contained, some mathematical techniques not usually covered in the undergraduate level are included in the appendix. At the end of each chapter, several practice problems are given.

There are six chapters in the book. Chapter 1 begins with the Maxwell equations in time-harmonic form. It is shown that, in a layered medium, the electromagnetic fields can be decomposed into two types of independent modes: the transverse magnetic (TM) mode and the transverse electric (TE) mode. In such a layered medium, the longitudinal component of the displacement vector alone can describe the TM mode. Similarly the longitudinal component of the magnetic displacement vector can describe the TE mode. The differential equations satisfied by these two displacement vectors are deduced. Finally the boundary conditions for electric and magnetic wall boundaries are discussed.

Chapter 2 presents a technique to find the TM and TE field components in a layered medium. In the spectral domain, the displacement vectors are shown to follow transmission line equations. Equivalent transmission line circuits for both modes are developed. The conventional transmission line theory is then applied to find the fields. Procedures for obtaining the field solutions for planar and nonplanar electric and magnetic current sources are outlined at the end.

Chapter 3 illustrates the applications of the technique developed in Chapter 2. Two canonical problems are considered for this purpose. The first problem is a magnetic line current over a grounded dielectric structure. The electromagnetic fields produced by this magnetic current source are obtained from the spectral domain transmission line equivalent circuit. Using the theory of complex integration, the total field is decomposed into space and surface wave fields. The radiation pattern of the source is obtained by invoking the saddle point method of integration. The space and surface wave power produced by the source are determined. The problem of electric line current source is solved in a similar manner. A procedure for finding fields produced by circular source is outlined at the end.

Chapter 4 presents important characteristics of the space and surface wave modes. It is shown that two surface wave modes are orthogonal to each other. Also, a surface wave mode and the space wave mode are orthogonal. The orthogonal properties together with reciprocity are utilized to find the unknown surface wave fields produced by an arbitrary source in a layered structure. Numerical results for space and surface wave power produced by a circular magnetic current source are shown.

In Chapter 5, analyses of printed radiating structures are presented. For printed antennas, two commonly used models are described: the edge admittance model and the patch current model. Spectral domain analysis is employed to determine the edge admittance of a circular patch antenna. The patch current model is demonstrated through an example of a slot coupled dipole. An analysis of infinite array of line sources over a dielectric coated ground plane is presented. Qualitative analysis of grating lobes, scan blindness and associated impedance mismatch problem are presented. The chapter ends with an analysis of a screen polarizer. The equivalent circuit of a typical screen polarizer is shown. The equivalent susceptance of a screen made of narrow thin conductive strips is deduced. Numerical results for a two-layered polarizer are shown.

Chapter 6 presents the analyses of planar transmission lines. They include the microstrip line, the coupled line, and the slot line. In the first part of the chapter, spectral domain static analysis of microstrip line and coupled lines are presented. The stationary expressions are used to find the static capacitance per unit length of the transmission lines. The characteristic impedance and the effective dielectric constant are expressed in terms of the capacitance. Numerical results for these two transmission lines are also shown. The full wave analysis of microstrip line and slot line are presented next. The full wave analysis reveals the dispersive nature of a transmission line. Variations of the characteristic impedance and effective dielectric constant with frequency are shown.

Appendix A1 describes the saddle point method of integration. Appendix A2 outlines the singularity extraction technique. Appendix A3 discusses the branch point and branch cut of a multivalued function. A procedure for integrating around a branch cut is also presented. A bibliography of books and articles for supplemental reading is given.

I would like to express my sincere gratitude to my teachers, who have been a constant inspiration. A major portion of the research work, included in this book, was conducted under the supervision of Ramesh Garg of IIT Kharagpur, India. I gratefully acknowledge his contribution and support. I am grateful to the reviewer for his many valuable comments and constructive suggestions, which helped to improve the quality of the book. I also appreciate the support from the staff members of Artech House during the entire project. Special thanks are due to my friends and colleagues in Hughes: Bruce Williams, Luciano Lopez, Louis Fermilia, Gordon Rollins and Alan Keith for their encouragement and time spent in proofreading the manuscript. The entire manuscript was typed by my wife Arundhuti. I am grateful to her for her invaluable contribution to the creation of the book. She was instrumental to the success of the project. I also thank my 5-year-old daughter, Atreyi, who sacrificed the company of her beloved parents for many weekends and evenings. Finally, this book is dedicated to my mother, who has been a constant support to me throughout my life.

Chapter 1
Field Equations in Layered Media

1.1 INTRODUCTION

In this chapter, governing equations for the electromagnetic fields in layered media are derived. It is shown that, outside the source region and inside a particular layer, the electromagnetic fields can be decomposed into two types of modes: the transverse magnetic (TM) and the transverse electric (TE) modes. The TM mode has no magnetic field component along a chosen reference direction; that is, the resultant magnetic field vector is transverse to the reference direction. For a homogeneous medium the reference direction can be selected arbitrarily. However, a layered structure as a whole, can support only a TM or TE mode for which the reference direction is along the normal of the planar interface of two dissimilar layers.

The organization of this chapter is as follows. Maxwell's equations and continuity equations in the "time-harmonic" forms are presented first. Next we show that, to determine the various components of the electric and magnetic field vectors for the TM mode, one needs to know only the longitudinal component of the electric displacement vector. In other words, the longitudinal component of the electric displacement vector alone provides the necessary and sufficient information about all other components of the TM mode fields. Similarly, for the TE mode, all components of the field vectors can be determined from the longitudinal component of the magnetic displacement vector. Because the complete field solution is the superposition of the TM and TE modes, the longitudinal components of electric and magnetic displacement vectors are the key quantities that can describe the complete field solution. The differential equations satisfied by these two quantities in a layered medium are derived. Finally, boundary conditions for electric and magnetic wall boundaries and the behavior at the interface of two layers are discussed. The inability to support other modes is discussed at the end of this chapter.

1.2 TIME-HARMONIC FORM OF MAXWELL'S EQUATIONS

In the time-harmonic form, the field quantities and source quantities, that is, the components of the field and source vectors, are represented by complex numbers.

The magnitude of a complex quantity is chosen to be the root-mean-square (r.m.s.) value of that quantity.[1] Also, we consider that the time dependent factor to be exp ($j\omega t$), where ω is the angular frequency. With this time factor, the time derivative $\partial/\partial t$ can be replaced by $j\omega$, so that Maxwell's equations take the following form:

$$\nabla \times \mathbf{E} = -j\omega\mathbf{B} - \mathbf{J}_m \tag{1.1a}$$

$$\nabla \times \mathbf{H} = j\omega\mathbf{D} + \mathbf{J}_e \tag{1.1b}$$

$$\nabla \cdot \mathbf{D} = \rho_e \tag{1.1c}$$

$$\nabla \cdot \mathbf{B} = \rho_m \tag{1.1d}$$

where

$\mathbf{E}$ = electric field vector (volts/meter)
$\mathbf{H}$ = magnetic field vector (amperes/meter)
$\mathbf{D}$ = electric displacement vector (coulombs/meter^2)
$\mathbf{B}$ = magnetic displacement vector (webers/meter^2)
$\mathbf{J}_e$ = electric current density vector (amperes/meter^2)
$\mathbf{J}_m$ = magnetic current density vector (volts/meter^2)
ρ_e = electric charge density (coulombs/meter^3)
ρ_m = magnetic charge density (webers/meter^3)

The displacement vectors **D** and **B** are related to the respective field intensity vectors, **E** and **H**. For isotopic media the relations are

$$\mathbf{D} = \varepsilon\mathbf{E} \tag{1.2a}$$

$$\mathbf{B} = \mu\mathbf{H} \tag{1.2b}$$

where ε and μ are the permittivity and permeability, respectively, of the medium in which the field exists. For a lossless medium they are real quantities. For a lossy medium, they are complex numbers, with the imaginary parts representing the loss. In an anisotropic medium, each component of **D** (or **B**) is linearly related to all three components of **E** (or **H**), and ε (or μ) is no longer a scalar quantity, but can be represented by a 3 × 3 matrix form. Such quantities are known as *tensors of rank 2* or dyadics. However, we will restrict ourselves to isotropic media, so that ε and

[1] In many other books, the magnitude represents the peak value, which is $\sqrt{2}$ times of the r.m.s. value.

μ are scalars. The charge densities are related to the current densities via the following continuity equations, which represent the principle of conservation of charge:

$$\nabla \cdot \mathbf{J}_e = -j\omega\rho_e \tag{1.3a}$$

$$\nabla \cdot \mathbf{J}_m = -j\omega\rho_m \tag{1.3b}$$

These relations can also be proven by taking the divergence of (1.1a) and (1.1b) and then using (1.1c) and (1.1d). Therefore, if the current densities are known, the charge densities can be determined.

1.3 TM AND TE FIELDS

A major objective is to develop a technique to solve Maxwell's equations in a planar layered media. The easiest way to handle such a multilayered boundary value problem, perhaps, would be to find solutions for the TE and TM modes, separately. The transverse plane may be considered as the plane parallel to the interface of two layers (Figure 1.1). This particular selection of the transverse plane (z-plane) will decouple the boundary conditions for the TE and TM waves. Furthermore, inside the layered structure shown in Figure 1.1, independent existence of either TE (to z) or TM (to z) waves is possible. This is not true, in general, for any arbitrary choice of transverse plane.[2]

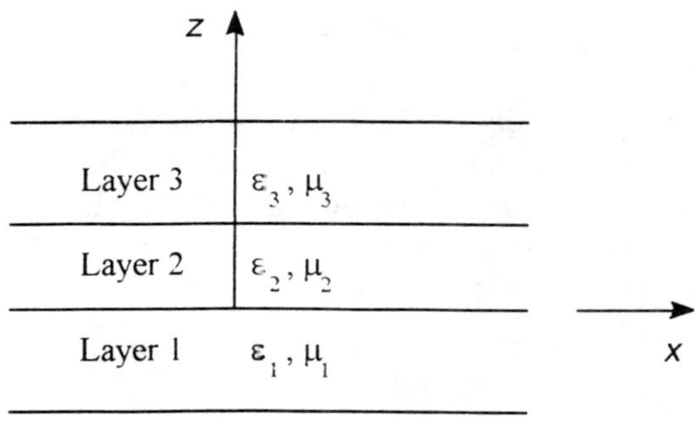

Figure 1.1 Geometry of a multilayered structure.

[2]A detailed explanation is given in Section 1.6.

In this section, we will show that, inside a source free homogeneous layer, the TM_z field[3] components can be determined uniquely if the longitudinal component (the z-component) of the electric displacement vector, D_z, is known. Similarly, the TE_z field components can be obtained if the longitudinal component of the magnetic displacement vector, B_z, is known.

By definition, the TM_z wave has no magnetic field component along z. For the TM_z wave, the source-free Maxwell's equations then become

$$\nabla \times \mathbf{E} = -j\omega \mathbf{B}_t \tag{1.4a}$$

$$\nabla \times \mathbf{H}_t = j\omega \mathbf{D} \tag{1.4b}$$

$$\nabla \cdot \mathbf{D} = 0 \tag{1.4c}$$

$$\nabla \cdot \mathbf{B}_t = 0 \tag{1.4d}$$

The subscript t in $\mathbf{B}$ and $\mathbf{H}$ indicates that these vectors lie on the transverse plane; that is, $\mathbf{B}_t \cdot \hat{z} = 0 = \mathbf{H}_t \cdot \hat{z}$. Because $\nabla \times \mathbf{B}_t = 0$, $\mathbf{B}_t$ can be expressed as the curl of another vector, $\mathbf{A}$. Symbolically,

$$\mathbf{B}_t = \nabla \times \mathbf{A} \tag{1.5}$$

Further, $\mathbf{B}_t$ has no $\hat{z}$ component. Therefore, one can choose vector $\mathbf{A}$ as

$$\mathbf{A} = \hat{z}A_z \tag{1.6}$$

based on the vector identity $\hat{z} \cdot \nabla \times (\hat{z}A_z) = 0$. We will now show that, in each homogeneous layer, A_z and D_z are directly related through a linear differential operator. Substituting (1.5) in (1.4b) and using $\mathbf{B}_t = \mu \mathbf{H}_t$, one obtains $1/\mu\ \nabla \times \nabla \times \mathbf{A} = j\omega \mathbf{D}$. Using the standard identity $\nabla \times \nabla \times \mathbf{A} = \nabla(\nabla \cdot \mathbf{A}) - \nabla^2 \mathbf{A}$, and taking the z-component in both sides, we get

$$\frac{1}{\mu}\left(\frac{\partial^2 A_z}{\partial z^2} - \nabla^2 A_z\right) = j\omega D_z \tag{1.7}$$

Again, it can be shown from (1.4) that $\mathbf{B}_t$ follows the wave equation in a homogeneous and source free layer; that is,

$$\nabla^2 \mathbf{B}_t + k^2 \mathbf{B}_t = 0, \qquad k^2 = \omega^2 \mu \varepsilon \tag{1.8}$$

[3]The subscript z denotes the preferred axis.

Taking the curl of (1.8) and changing the sequence of the operators, one can find that A_z also follows the same equation as $\mathbf{B}_t$. Therefore, $\nabla^2 A_z = -k^2 A_z$. With this substitution, (1.7) becomes

$$\frac{1}{\mu}\left(\frac{\partial^2}{\partial z^2} + k^2\right)A_z = j\omega D_z \tag{1.9}$$

or

$$A_z = j\omega\mu\left(\frac{\partial^2}{\partial z^2} + k^2\right)^{-1} D_z$$

Therefore,

$$\mathbf{A} = \hat{z}A_z = \hat{z}j\omega\mu\left(\frac{\partial^2}{\partial z^2} + k^2\right)^{-1} D_z \tag{1.10}$$

Equation (1.10) is an equivalent representation of (1.9) in terms of an inverse operator. To obtain A_z for a given D_z, one must solve (1.9) directly. For special cases, where D_z has cosine or sine variation along z, one can use (1.10). In such cases, the inverse operator becomes equivalent to a constant coefficient, which is independent of z. Equation (1.10) shows the direct relation between $\mathbf{A}$ and D_z. Note that the vector $\mathbf{A}$ can be determined from D_z only. Once $\mathbf{A}$ is known, $\mathbf{B}_t$ can be obtained from (1.5).

Using (1.4b) we can determine the electric field vector, $\mathbf{E}$. The TM_z field components are expressed in terms of A_z:

$$B_x = \frac{\partial A_z}{\partial y} \qquad B_y = \frac{-\partial A_z}{\partial x} \qquad B_z = 0$$
$$D_x = \frac{1}{j\omega\mu}\frac{\partial^2 A_z}{\partial x \partial z} \qquad D_y = \frac{1}{j\omega\mu}\frac{\partial^2 A_z}{\partial y \partial z} \tag{1.11}$$

Following a similar procedure, one can prove that the TE_z field component can be expressed in terms of another vector function $\mathbf{F} = \hat{z}F_z$, where F_z is functionally related to B_z alone. The various field components, in terms of B_z, can be derived as

$$B_x = \frac{1}{j\omega\varepsilon}\frac{\partial^2 F_z}{\partial x \partial z} \qquad B_y = \frac{1}{j\omega\varepsilon}\frac{\partial^2 F_z}{\partial y \partial z}$$
$$D_x = \frac{-\partial F_z}{\partial y} \qquad D_y = \frac{\partial F_z}{\partial x} \qquad D_z = 0 \tag{1.12}$$

where

$$F_z = j\omega\varepsilon \left(\frac{\partial^2}{\partial z^2} + k^2 \right)^{-1} B_z$$

We have seen that inside a source-free homogeneous layer (region), the electromagnetic fields can be determined from the longitudinal components of the electric and magnetic displacement vectors. The complexity, however, arises in the source region, which was so far excluded from consideration. Unfortunately, to determine the total complex power radiated by a source, one usually[4] needs to calculate the appropriate type of field (for a magnetic current source, the field type is "magnetic," and for an electric current source, the field type is "electric"[5]) in the source region. To circumvent this problem, one may consider a fictitious infinitely thin layer, which contains the observation point. Because the layer is very thin, the amount of source confined in the layer is negligibly small, and the layer can be regarded as source free.

For some special cases, however, the source density may be infinitely large at a given point or set of points, so that the fictitious layer would include a finite amount of source, which cannot be ignored. Examples of these sources are a point source, a line current source, and a surface current source. For the first two sources (that is, the point source and the line current source) the required field intensities at the source points are singular (infinitely large) in nature, which makes the complex power infinitely large. These types of sources can never be realized in practice. The surface current sources, however, exist in practice; for example, the induced current on a perfect conducting sheet or equivalent currents on the aperture surface of a radiating aperture. For a surface current the fields in the source region are finite. Moreover, the field type required for complex power calculation is also continuous across the source surface. The continuity property of the required field is generally utilized, and one calculates the field intensity at a source-free location that is at an infinitesimal distance from the source surface.

1.4 GOVERNING EQUATION FOR THE TM$_z$ FIELDS

In the preceding section, we showed that from the longitudinal components of the electric and magnetic displacement vectors (D_z, B_z) one can construct the complete

[4]For a homogeneous medium, the complex power can also be calculated by integrating the Poynting vector in the visible and invisible regions. See D.R. Rhodes, "On a Fundamental Principle in the Theory of Planar Antennas," *Proceedings of IEEE*, Sept. 1964, pp. 1013–1921. For a nonhomogeneous structure, this method is not straightforward.

[5]For a magnetic current source, the complex power is $P_m = -\int \mathbf{J}_m \cdot \mathbf{H}^* \, dv$. For an electric current source, the complex power is $P_e = -\int \mathbf{J}_e^* \cdot \mathbf{E} \, dv$. The integrations are taken over the source volume. The asterisk sign denotes the complex conjugate. See R. F. Harrington, *Time-Harmonic Electromagnetic Fields*, McGraw-Hill, New York, 1961, p. 96, 1961.

field solution inside a homogeneous layer. One therefore, should find these two quantities: D_z and B_z. Inside a particular source-free homogeneous layer, D_z satisfies the wave equation $\nabla^2 D_z + k^2 D_z = 0$. This equation, however, is not valid at the interface of two dissimilar layers. In this section, we deduce an equation for D_z that is valid everywhere inside a multilayered structure, in general a nonhomogeneous medium.

Let **D** be the electric displacement vector, produced by electric and magnetic current sources, $\mathbf{J}_e$ and $\mathbf{J}_m$, respectively. Using a vector identity and (1.1a), we can express $\nabla \times \mathbf{D}$ as

$$\nabla \times \mathbf{D} = \nabla \times (\varepsilon \mathbf{E}) = \varepsilon \nabla \times \mathbf{E} + \nabla \varepsilon \times \mathbf{E} = \varepsilon(-\mathbf{J}_m - j\omega\mu\mathbf{H}) + \frac{\partial \varepsilon}{\partial z}\hat{z} \times \mathbf{E} \quad (1.13)$$

Note that we consider the permittivity of the medium to be a function of z, which is true for a nonhomogeneous medium whose inhomogeneity is along z, such as the medium of Figure 1.1. For a layered structure, $\varepsilon(z)$ can be represented by a series of step functions. Similarly, the permeability is also a function of z, in general. Taking curl of (1.13) one obtains

$$\nabla \times \nabla \times \mathbf{D} = -\nabla \times (\varepsilon \mathbf{J}_m) - j\omega \nabla \times (\mu\varepsilon\mathbf{H}) + \nabla \times \left[\frac{\partial \varepsilon}{\partial z}(\hat{z} \times \mathbf{E})\right] \quad (1.14)$$

The lefthand side of (1.14) can be expressed as $\nabla(\nabla \cdot \mathbf{D}) - \nabla^2 \mathbf{D}$. Using the vector identity $\nabla \times (\phi \mathbf{A}) = \nabla\phi \times \mathbf{A} + \phi \nabla \times \mathbf{A}$, where ϕ is a scalar function, in the last two terms of (1.14), we can write

$$\nabla(\nabla \cdot \mathbf{D}) - \nabla^2 \mathbf{D} = -\nabla \times (\varepsilon \mathbf{J}_m) - j\omega\mu\varepsilon \nabla \times \mathbf{H} - j\omega \frac{\partial}{\partial z}(\mu\varepsilon)\, \hat{z} \times \mathbf{H}$$

$$+ \frac{\partial^2 \varepsilon}{\partial z^2}\hat{z} \times (\hat{z} \times \mathbf{E}) + \frac{\partial \varepsilon}{\partial z} \nabla \times (\hat{z} \times \mathbf{E}) \quad (1.15)$$

Using Maxwell's equations (1.1b) and (1.1c) and retaining the z component terms only, one obtains

$$\frac{\partial \rho_e}{\partial z} - \nabla^2 D_z = -\hat{z} \cdot \nabla \times (\varepsilon \mathbf{J}_m) - j\omega\mu\varepsilon\, (\mathbf{J}_e \cdot \hat{z}) + k^2 D_z + \frac{\partial \varepsilon}{\partial z}\hat{z} \cdot [\nabla \times (\hat{z} \times \mathbf{E})] \quad (1.16)$$

where $k^2 = \omega^2 \mu(z)\varepsilon(z)$. Again, it can be shown that $\hat{z} \cdot \nabla \times (\hat{z} \times \mathbf{E}) = \nabla_t \cdot \mathbf{E}_t$, where

$$\nabla_t \cdot E_t = \frac{\partial E_x}{\partial x} + \frac{\partial E_y}{\partial y}$$

Using equation (1.1c), one can express $\hat{z} \cdot \nabla \times (\hat{z} \times \mathbf{E})$ as

$$\hat{z} \cdot \nabla \times (\hat{z} \times \mathbf{E}) = \nabla_t \cdot \mathbf{E}_t = -\frac{1}{\varepsilon}\frac{\partial D_z}{\partial z} + \frac{\nabla \cdot \mathbf{D}}{\varepsilon} = \frac{\rho_e}{\varepsilon} - \frac{1}{\varepsilon}\frac{\partial D_z}{\partial z} \tag{1.17}$$

Substituting (1.17) in (1.16), we finally derive

$$\nabla^2 D_z + k^2 D_z - \frac{\partial(\ln\varepsilon)}{\partial z}\frac{\partial D_z}{\partial z} = \frac{\partial \rho_e}{\partial z} - \frac{\partial(\ln\varepsilon)}{\partial z}\rho_e + \hat{z} \cdot \nabla \times (\varepsilon \mathbf{J}_m) + j\omega\mu\varepsilon(\hat{z} \cdot \mathbf{J}_e) \tag{1.18}$$

Note that the only unknown quantity in (1.18) is D_z. The solution procedure will be discussed in the next chapter. A very important feature of (1.18) is that it yields the necessary source condition for the TM_z field generation. If the sources are distributed in such a way that the righthand side of (1.18) vanishes everywhere in the source region, then these sources will not be able to produce any TM_z field. Mathematically, if the TM_z field is not to be generated by a source, then the following condition must hold everywhere inside the source volume:

$$\frac{\partial \rho_e}{\partial z} - \rho_e\frac{\partial(\ln\varepsilon)}{\partial z} + \hat{z} \cdot \nabla \times (\varepsilon \mathbf{J}_m) + j\omega\mu\varepsilon(\hat{z} \cdot \mathbf{J}_e) = 0 \tag{1.19}$$

If this condition is not satisfied, the source will produce the TM_z fields.

1.4.1 Boundary Conditions for $\mathbf{D}_z$

To solve the partial differential equation (1.18), the boundary conditions for D_z should be specified. In a layered structure, three different boundaries may occur: (i) electric wall, (ii) magnetic wall, and (iii) interface of two layers.[6] We shall consider them one at a time.

Let $\hat{n}$ be the unit normal vector on a boundary surface. For an electric wall boundary, the tangential electric field must vanish, so that

$$\hat{n} \times \mathbf{D} = 0 \tag{1.20}$$

[6]The interface boundary condition is not required for solving (1.18), because the equation is valid everywhere in the layered medium including the interface.

If the electric wall is parallel to the z-axis, then $\hat{n}$ will be perpendicular to $\hat{z}$, and (1.20) yields

$$D_z = 0 \tag{1.21}$$

This boundary condition is known as the *homogeneous Dirichlet boundary condition*. If the electric wall is perpendicular to the z-axis, then $\hat{n} = \hat{z}$. In such a case, (1.20) cannot provide any condition for D_z. To obtain the required boundary condition, we start with the following Maxwell's equation:

$$\nabla \cdot \mathbf{D} = 0 \tag{1.22}$$

(We do not consider any source on a boundary surface.)

Equation (1.22) can alternatively be written as

$$\varepsilon(\nabla_t \cdot \mathbf{E}_t) + \frac{\partial D_z}{\partial z} = 0 \tag{1.23}$$

Because $\mathbf{E}_t = 0$ on the electric wall surface, we have

$$\frac{\partial D_z}{\partial z} = 0 \tag{1.24}$$

This condition is known as the *homogeneous Neumann boundary condition*. The magnetic wall boundary conditions are different from the electric wall boundary conditions. When the magnetic wall is parallel to z, we obtain the Neumann condition for D_z; that is,

$$\frac{\partial D_z}{\partial n} = 0 \tag{1.25}$$

where $\partial D_z/\partial n$ represents the directional derivative of D_z along the normal direction of the magnetic wall. This result can be proven in the following way. Without loss of generality, one can assume $\hat{n} = \hat{y}$, so that $H_x = H_z = E_y = 0$ on the magnetic wall. Now using Maxwell's equation $\nabla \times \mathbf{E} = j\omega\mu\mathbf{H}$, one obtains $\partial E_z/\partial y = 0$, which yields $\partial D_z/\partial n = 0$. If the magnetic wall is perpendicular to the z-axis, that is, parallel to the interface of two consecutive layers, we obtain the Dirichlet boundary condition:

$$D_z = 0 \tag{1.26}$$

At the interface of two layers (Figure 1.1) whose electrical parameters are (ε_1, μ_1) and (ε_2, μ_2) respectively, the normal component of the electrical displacement vector must be continuous. That is,

$$D_z^{(1)} = D_z^{(2)} \tag{1.27}$$

where the subscripts (1) and (2) denote the quantity in layer 1 and 2, respectively. The electric field components tangential to the interface must be continuous also, that is, $\mathbf{E}_t^{(1)} = \mathbf{E}_t^{(2)}$. Applying (1.23) for each layer, we can show

$$\frac{1}{\varepsilon_1}\frac{\partial D_z^{(1)}}{\partial z} = \frac{1}{\varepsilon_2}\frac{\partial D_z^{(2)}}{\partial z} \tag{1.28}$$

1.5 GOVERNING EQUATION FOR THE TE$_z$ FIELDS

In a homogeneous source-free layer, the TE$_z$ field components can be determined from the longitudinal component of the magnetic displacement vector, B_z. The differential equation satisfied by B_z in a nonhomogeneous medium (inhomogeneity along z) can be obtained from Maxwell's equation. The procedure is the same as that for D_z, which is detailed in Section 1.4. The final form is

$$\nabla^2 B_z + k^2 B_z - \frac{\partial B_z}{\partial z}\left[\frac{\partial(\ln\mu)}{\partial z}\right] = \frac{\partial \rho_m}{\partial z} - \rho_m \frac{\partial(\ln\mu)}{\partial z} - \hat{z}\cdot\nabla\times(\mu\mathbf{J}_e) + j\omega\mu\varepsilon(\mathbf{J}_m\cdot\hat{z}) \tag{1.29}$$

If the source distribution is such that everywhere in the source volume the following relation holds

$$\frac{\partial \rho_m}{\partial z} - \rho_m \frac{\partial(\ln\mu)}{\partial z} - \hat{z}\cdot\nabla\times(\mu\mathbf{J}_e) + j\omega\mu\varepsilon(\mathbf{J}_m\cdot\hat{z}) = 0 \tag{1.30}$$

then this particular source will not produce any TE$_z$ field.

1.5.1 Boundary Conditions for B_z

The boundary conditions for B_z on the electric and magnetic walls can be derived in the same manner as it was derived for D_z. The boundary conditions of a magnetic field on a magnetic wall are identical to that of an electric field on an electric wall. Therefore, from duality and the boundary conditions for D_z, we directly write the following boundary conditions for B_z.

Magnetic wall parallel to z-axis:

$$B_z = 0 \tag{1.31}$$

Magnetic wall parallel to the interface:

$$\frac{\partial B_z}{\partial z} = 0 \tag{1.32}$$

Electric wall parallel to z-axis:

$$\frac{\partial B_z}{\partial n} = 0 \tag{1.33}$$

Electric wall parallel to the interface of two layers:

$$B_z = 0 \tag{1.34}$$

At the interface of two layers, any discontinuity in the magnetic field occurs only if the permeabilities of the associated layers differ from one another. Therefore, at the interface, the boundary conditions for B_z will be the same as that for D_z, except ε will be replaced by μ. Recalling the interface boundary conditions for D_z, (1.27) and (1.28), we can have the following boundary conditions for B_z:

$$B_z^{(1)} = B_z^{(2)} \tag{1.35}$$

$$\frac{1}{\mu_1}\frac{\partial B_z^{(1)}}{\partial z} = \frac{1}{\mu_2}\frac{\partial B_z^{(2)}}{\partial z} \tag{1.36}$$

where $B_z^{(1)}$ and $B_z^{(2)}$ represent the longitudinal components of the magnetic displaced vectors in layer 1 and layer 2, respectively.

1.6 STRUCTURAL INABILITY TO SUPPORT AN ISOLATED TE$_x$, TM$_x$, TE$_y$, or TM$_y$ FIELD

In Section 1.3, we showed that, in a source-free homogeneous layer, the TM$_z$ and TE$_z$ field components can be determined from D_z and B_z, respectively. In Sections 1.4 and 1.5, equations satisfied by D_z and B_z were derived. Source conditions for the nonexistence of the TM$_z$ and TM$_z$ modes were also obtained (see conditions (1.19) and (1.30)). These two conditions are mutually independent. In other words, a source may satisfy (1.19) without satisfying (1.30), because neither the divergence nor the curl of a vector function can uniquely determine the vector function itself. The charge density is proportional to the divergence of the current density, so one can conclude by inspection that the lefthand sides of (1.19) and (1.30) are independent from one another. Equation (1.30) cannot be derived from (1.19) and vice versa. Therefore, it is possible to have a source distribution that satisfies (1.19) but not (1.30). This implies that the TE_z mode can be generated without exciting the TM_z mode. That is, the TM_z and TE_z modes can exist independently.

This is, however, not the case for any other mode. Except for some special conditions, the structure in Figure 1.1 is incapable of supporting any other mode independently. We intend to prove this for one case, for instance, the TE_x mode case. We will show that, in the source-free region, the TE_x mode cannot exist inside the layered structure shown in Figure 1.1. We now recall (1.15). Under the source-free condition, ($\rho_e = \rho_m = \mathbf{J}_e = \mathbf{J}_m = 0$), the equation becomes

$$\nabla(\nabla \cdot \mathbf{D}) - \nabla^2 \mathbf{D} = -j\omega\mu\varepsilon\nabla \times \mathbf{H} - j\omega \frac{\partial}{\partial z}(\mu\varepsilon)\hat{z} \times \mathbf{H} + \frac{\partial \varepsilon}{\partial z} \nabla \times (\hat{z} \times \mathbf{E}) - \frac{\partial^2 \varepsilon}{\partial z^2} \mathbf{E}_t$$

Substituting $\nabla \cdot \mathbf{D} = 0$, $\nabla \times \mathbf{H} = j\omega\mathbf{D}$, one obtains

$$\nabla^2 \mathbf{D} + k^2 \mathbf{D} = j\omega \frac{\partial}{\partial z}(\mu\varepsilon)\hat{z} \times \mathbf{H} - \frac{\partial \varepsilon}{\partial z} \nabla \times (\hat{z} \times \mathbf{E}) + \frac{\partial^2 \varepsilon}{\partial z^2} \mathbf{E}_t \tag{1.37}$$

Taking the x component only,

$$\nabla^2 D_x + k^2 D_x = -j\omega \frac{\partial}{\partial z}(\mu\varepsilon)H_y + \frac{\partial \varepsilon}{\partial z}\frac{\partial E_x}{\partial z} + \frac{\partial^2 \varepsilon}{\partial z^2} E_x \tag{1.38}$$

For the TE_x mode, $E_x = D_x = 0$. Using this in (1.38), one obtains

$$\frac{\partial}{\partial z}(\mu\varepsilon)H_y = 0 \tag{1.39}$$

This condition is satisfied if either $H_y = 0$ or $(\partial/\partial z)(\mu\varepsilon) = 0$. If $H_y = 0$, from Maxwell's equation, we get

$$E_x = \frac{1}{j\omega\mu\varepsilon}\frac{\partial B_z}{\partial y}$$

Because B_z is continuous across the interface of two layers, E_x has to be discontinuous due to the factor $1/\mu\varepsilon$. Now, E_x cannot be discontinuous at the interface. Therefore, for the TE_x mode to exist the second condition must be satisfied, that is, $(\partial/\partial z)(\mu\varepsilon) = 0$, or $\mu\varepsilon$ has to be independent of z, which is a very special case and not true in general. Therefore, we can conclude that a layered structure (layered along z) cannot support the TE_x mode. From the symmetry point of view, the structure would not support the TE_y mode either. In a similar fashion, one can prove that an isolated TM_x or TM_y mode cannot exist inside the structure.

PROBLEMS

1.1 If **A** is a differentiable vector function, show that $\hat{t} \cdot [\nabla \times (\hat{t} \times \mathbf{A})] = \nabla \cdot \mathbf{A} - \nabla(\mathbf{A} \cdot \hat{t}) \cdot \hat{t}$, where $\hat{t}$ is a unit vector along a given direction. In particular, when $\hat{t} = \hat{z}$, show that the righthand side simplifies to $\partial A_x/\partial x + \partial A_y/\partial y$.

1.2 Show that in a source-free region, the TE_z field components are obtainable from the z component of the magnetic vector potential.

1.3 Derive (1.29). Obtain the boundary conditions for three different boundaries (magnetic wall, electric wall, and interface of two media).

1.4 Show that a z-directed electric current source does not produce the TE_z field. Find the condition for which the produced field is TEM_z type. (Hint: use (1.19) and (1.30).)

1.5 Verify that the electric current corresponding to the TE_{mn} mode propagation that exists on the walls of a perfectly conducting rectangular wave guide satisfies (1.19).

1.6 Verify that the electric current on the perfectly conducting walls of a circular wave guide supporting the TM_{mn} mode satisfies (1.30).

1.7 A uniform electric line current along the $\hat{x}$ direction is placed in a nonhomogeneous medium, whose inhomogeneity is along $\hat{z}$. Show that this current will produce the TE_z mode fields only. Also show that an x-directed uniform magnetic line current will not produce the TE_z mode.

1.8 A circular magnetic line current is placed inside a nonhomogeneous (inhomogeneity along $\hat{z}$) medium. The current lies on the x-y plane. Assume that the current distribution is $f(\phi)$, where ϕ is the angular coordinate. Find $f(\phi)$ for which the TE_z mode will not be generated.

1.9 Show that a partially filled rectangular wave guide cannot support an isolated TE_z or TM_z mode (the direction of propagation is along z).

1.10 Show that, except for rationally symmetric field, a partially filled circular wave guide cannot support an isolated TE_z or TM_z mode, where z is the direction of propagation.

1.11 Show that a cylindrically layered structure, in general, does not support an isolated mode.

1.12 Show that a spherically layered structure supports an isolated TE_r or TM_r mode, where r represents the radial coordinate.

1.13 Prove that, in a spherically layered structure, the radial component of the electric displacement vector D_r satisfies the following differential equation:

$$\nabla^2 D_r + \left\{k^2 + \frac{2}{r^2} - \frac{2}{\varepsilon r}\frac{\partial \varepsilon}{\partial r}\right\} D_r + \left\{\frac{2}{r} - \frac{1}{\varepsilon}\frac{\partial \varepsilon}{\partial r}\right\}\frac{\partial D_r}{\partial r}$$

$$= \{\nabla \times (\varepsilon \mathbf{J}_m)\} \cdot \hat{r} + j\omega\mu\varepsilon(\mathbf{J}_e \cdot \hat{r}) + \frac{1}{r^2}\frac{\partial}{\partial r}(r^2 \rho_e) - \frac{1}{\varepsilon}\frac{\partial \varepsilon}{\partial r}\rho_e$$

1.14 Derive the differential equation for the radial component of the magnetic displacement vector, B_r, in a spherically layered structure.

1.15 A slot length l and width t is cut on an infinite ground plane at $z = 0$. The electric field distribution on the slot surface is approximated as

$$E_t = E_0 \cos\left(\frac{\pi x}{l}\right)$$

where E_t denotes the field component along the width of the slot. Show that both the TE_z and TM_z fields will be produced by the slot.

Chapter 2
Solutions for Field Equations

2.1 INTRODUCTION

This chapter presents a technique for finding unknown electromagnetic fields in layered media. Figure 2.1 shows the structure of a layered medium. In the preceding chapter we saw that to obtain the complete field solution, we need to know the longitudinal components of the electric and magnetic displacement vectors, D_z and B_z. To determine D_z and B_z we express the field quantities in terms of their two-dimensional Fourier spectra. In the spectral domain, the quantities D_z and B_z follow transmission line equations. The transmission line parameters are functions of the electrical parameters of a medium. The conventional transmission line theory is then used to obtain the Fourier spectra for D_z and B_z. For various planar excitation sources, expressions for D_z and B_z and other field quantities are presented. A procedure for nonplanar excitation sources is also outlined.

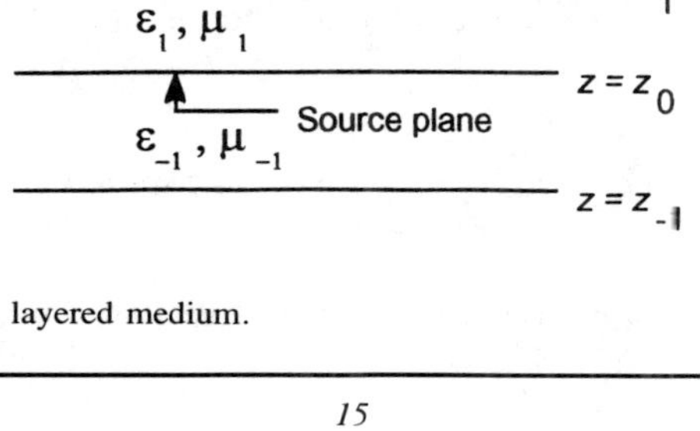

Figure 2.1 Geometry of a layered medium.

2.2 THE TM_z FIELD PRODUCED BY PLANAR ELECTRIC CURRENT

In this section we consider a planar electric current as a source for the TM_z fields. The planar electric current model has several useful applications in the areas of microstrip lines and antennas. In the moment method formulation, the top conductor of a microstrip antenna is usually replaced by a planar electric current source that contributes to the radiated field. In the case of a microstrip line, the surface current on the top conductor determines the guided field distribution. If the conductor is very thin the surface current that flows around the conductor is approximated as a planar current. Detailed analyses of microstrip antenna and microstrip transmission line are presented in Chapters 5 and 6, respectively.

2.2.1 The TM_z Field Equation in the Fourier Domain

The main objective is to find a solution for the TM_z field components, produced by a planar electric current source. The source plane is parallel to the *x-y* plane, as shown in Figure 2.1. The source is placed inside a layered medium (layered along *z* direction). To obtain all field components associated with the TM_z mode, we determine D_z first. For a planar electric current source (that is, when $\mathbf{J}_m = \mathbf{J}_e \cdot \hat{z} = 0$) D_z will satisfy the following differential equation (see (1.18)):

$$\nabla^2 D_z + k^2 D_z - \frac{\partial(\ln\varepsilon)}{\partial z}\frac{\partial D_z}{\partial z} = \frac{\partial \rho_e}{\partial z} - \frac{\partial(\ln\varepsilon)}{\partial z}\rho_e \tag{2.1}$$

Recall that ε is the permittivity of the medium and ρ_e is the electric charge density. Now, k is the wave number inside the structure, which is given by

$$k = \omega\sqrt{\mu\varepsilon}$$

where ω is the angular frequency. Because μ (the permeability) and ε are functions of z, k is also a function of z. The simplest way to solve this partial differential equation, perhaps, is to solve in the Fourier domain. We define a two-dimensional Fourier transform (or Fourier spectrum) of a function $f(x, y)$ as

$$F\{f(x, y)\} = \tilde{f}(k_x, k_y) = \frac{1}{4\pi^2}\int_x\int_y f(x, y)\exp(jk_x x + jk_y y)\,dx\,dy \tag{2.2a}$$

The function $f(x, y)$ can be expressed in terms of its two-dimensional Fourier transform as

$$f(x, y) = \int_{-\infty}^{\infty}\int_{-\infty}^{\infty} \tilde{f}(k_x, k_y) \exp(-jk_x x - jk_y y)\, dk_x\, dk_y \tag{2.2b}$$

Unless otherwise mentioned, we will use this definition of Fourier transform throughout the book. Using this definition, we express D_z as

$$D_z(x, y, z) = \int_{-\infty}^{\infty}\int_{-\infty}^{\infty} \tilde{D}_z(k_x, k_y, z) \exp(-jk_x x - jk_y y)\, dk_x\, dk_y \tag{2.3a}$$

where D_z and $\tilde{D}_z$ are Fourier transform pairs, so that

$$\tilde{D}_z(k_x, k_y, z) = \frac{1}{4\pi^2}\int_{-\infty}^{\infty}\int_{-\infty}^{\infty} D_z\,(x, y, z) \exp(jk_x x + jk_y y)\, dx\, dy \tag{2.3b}$$

Substituting (2.3a) in (2.1) and changing the sequence of the differential and integral operators, we derive,

$$\int_{-\infty}^{\infty}\int_{-\infty}^{\infty} \left\{\frac{\partial^2 \tilde{D}_z}{\partial z^2} + k_z^2\, \tilde{D}_z\right\} \exp(-jk_x x - jk_y y)\, dk_x\, dk_y$$

$$= \frac{\partial\, \ln\varepsilon}{\partial z}\int_{-\infty}^{\infty}\int_{-\infty}^{\infty} \frac{\partial \tilde{D}_z}{\partial z} \exp(-jk_x x - jk_y y)\, dk_x\, dk_y + \frac{\partial \rho_e}{\partial z} - \rho_e \frac{\partial\, \ln\varepsilon}{\partial z} \tag{2.4}$$

where $k_z^2 = k^2 - k_x^2 - k_y^2$. Taking Fourier inverse in both sides, we have

$$\frac{\partial^2 \tilde{D}_z}{\partial z^2} + k_z^2 \tilde{D}_z = \frac{\partial \tilde{D}_z}{\partial z}\frac{\partial\, \ln\varepsilon}{\partial z} + \frac{1}{4\pi^2}\int_{-\infty}^{\infty}\int_{-\infty}^{\infty}\left[\frac{\partial \rho_e}{\partial z} - \rho_e \frac{\partial\, \ln\varepsilon}{\partial z}\right] \exp(jk_x x + jk_y y)\, dx\, dy \tag{2.5}$$

Now, from (2.2), the two-dimensional Fourier transform of ρ_e is given by

$$\tilde{\rho}_e(k_x, k_y, z) = \frac{1}{4\pi^2}\int_{-\infty}^{\infty}\int_{-\infty}^{\infty} \rho_e \exp(jk_x x + jk_y y)\, dx\, dy$$

Therefore, (2.5) becomes

$$\frac{\partial^2 \tilde{D}_z}{\partial z^2} + k_z^2 \tilde{D}_z = \frac{\partial \tilde{D}_z}{\partial z}\frac{\partial\, \ln\varepsilon}{\partial z} + \frac{\partial \tilde{\rho}_e}{\partial z} - \tilde{\rho}_e \frac{\partial\, \ln\varepsilon}{\partial z} \tag{2.6}$$

Equation (2.6) is the Fourier domain representation of (2.1).

2.2.2 The Transmission Line Analogy

Equation (2.6) is, in general, a second-order nonhomogeneous linear differential equation. The inhomogeneity is due to the factors $(\partial \ln\varepsilon/\partial z)$ and k_z^2, which are functions of z. In a layered structure, ε and k_z can be represented in terms of step functions. Because (2.6) is nonhomogeneous, it is difficult to obtain a closed form expression for $\tilde{D}_z$. Moreover, it is difficult to solve $\tilde{D}_z$ directly from (2.6), although, in principle, it is possible.

The difficulty of solving (2.6) is circumvented by invoking the transmission line analogy. By means of Maxwell's equations, we reduce (2.6) in the form of telegrapher's equations pair. This enables us to determine the solution for $\tilde{D}_z$ almost by inspection, especially for a planar source of excitation. Toward that end, we recall the following Maxwell's equation:

$$\nabla \cdot \mathbf{D} = \rho_e \tag{2.7}$$

Taking the Fourier transformation of (2.7), we write

$$\frac{\partial \tilde{D}_z}{\partial z} + \varepsilon \tilde{I} = \tilde{\rho}_e \tag{2.8}$$

In this equation, $\tilde{I}$ represents the Fourier transformation of $\nabla_t \cdot \mathbf{E}_t$ that is equal to $\partial E_x/\partial x + \partial E_y/\partial y$. Therefore,

$$\tilde{I}(k_x, k_y, z) = \frac{1}{4\pi^2} \int_{-\infty}^{\infty} \int_{-\infty}^{\infty} \nabla_t \cdot \mathbf{E}_t \exp(jk_x x + jk_y y)\, dx\, dy \tag{2.9}$$

Differentiating (2.8) with respect to z we obtain

$$\frac{\partial^2 \tilde{D}_z}{\partial z^2} + \varepsilon \frac{\partial \tilde{I}}{\partial z} + \tilde{I} \frac{\partial \varepsilon}{\partial z} = \frac{\partial \tilde{\rho}_e}{\partial z} \tag{2.10}$$

Combining (2.6) and (2.10) we can eliminate the second-order derivative terms. The result is

$$k_z^2 \tilde{D}_z - \varepsilon \frac{\partial \tilde{I}}{\partial z} - \tilde{I} \frac{\partial \varepsilon}{\partial z} = \frac{\partial \tilde{D}_z}{\partial z} \frac{\partial \ln\varepsilon}{\partial z} - \tilde{\rho}_e \frac{\partial \ln\varepsilon}{\partial z} \tag{2.11}$$

Equations (2.11) and (2.8) can be combined together to yield

$$\frac{\partial \tilde{I}}{\partial z} = \frac{k_z^2}{\varepsilon} \tilde{D}_z \tag{2.12}$$

Equations (2.8) and (2.12) are telegrapher's equations pair, where $\tilde{D}_z$ and $\tilde{I}$ can be regarded as the line voltage and the line current, respectively. The transmission line parameters would be series impedance[1] per unit length $z_s = \varepsilon$ and shunt admittance per unit length $y_s = -k_z^2/\varepsilon$. The characteristic admittance of the transmission line is given by

$$y_0^m = \sqrt{\frac{y_s}{z_s}} = \frac{jk_z}{\varepsilon} \tag{2.13}$$

where the superscript corresponds to the TM$_z$ (transverse *magnetic*, *m* stands for magnetic) mode. The propagation constant of the transmission line is

$$\beta = \sqrt{-y_s z_s} = k_z \tag{2.14}$$

When k_z^2 is negative, β is imaginary. To ensure that the field decays along the positive direction of z, we take the negative imaginary root for β. Mathematically,

$$\beta = k_z \tag{2.15a}$$

when k_z is real and

$$\beta = -j\,|k_z| \tag{2.15}$$

when k_z is imaginary. For a given k_x and k_y, β and y_0^m are different in different layers. A layered structure is therefore equivalent to transmission line sections connected in cascade. The transmission line length of a section is equal to the thickness of the associated layer.

Interestingly, a necessary condition for considering $\tilde{D}_z$ and $\tilde{I}$ as the line voltage and the line current is satisfied. For series-connected (in cascade) transmission line sections, the line voltage and line current are continuous across a junction. This ensures that D_z and $\mathbf{E}_t$ are continuous at the interface of two layers.

2.2.3 Solution for $\tilde{D}_z$

In the preceding section we found that $\tilde{D}_z$ and $\tilde{I}$ satisfy the following telegrapher's equations:

$$\frac{\partial \tilde{D}_z}{\partial z} + \varepsilon \tilde{I} = \tilde{\rho}_e \tag{2.16}$$

---

[1]Note the unusual units for the series impedance and shunt admittance. This is because the line voltage and line current do not have the units of voltage and current, respectively.

$$\frac{\partial \tilde{I}}{\partial z} - \frac{k_z^2}{\varepsilon}\tilde{D}_z = 0 \tag{2.17}$$

where $\tilde{I}$ and $\tilde{\rho}_e$ are the Fourier transforms of $\nabla_t \cdot \mathbf{E}_t$ and the charge density ρ_e, respectively. Further, $\tilde{D}_z$ and $\tilde{I}$ satisfy the continuity conditions at the junction of two adjoining transmission line sections. Therefore the transmission line theory is applicable for solving $\tilde{D}_z$ and $\tilde{I}$.

For a planar electric current source, that is, if the source lies on a plane surface, the current density vector, $\mathbf{J}_e$, can be expressed in terms of the Dirac-delta function. The charge density can also be expressed in the same way. Suppose ρ_{se} be the surface charge density (coulombs/m^2) on the $z = z_0$ plane. Then,

$$\rho_e = \rho_{se}\delta(z - z_0) \tag{2.18}$$

where $\delta(z - z_0)$ is the Dirac-delta function. In the Fourier domain, (2.18) becomes

$$\tilde{\rho}_e = \tilde{\rho}_{se}\delta(z - z_0) \tag{2.19}$$

Substituting the expression for $\tilde{\rho}_e$ in (2.16) one obtains

$$\frac{\partial \tilde{D}_z}{\partial z} + \varepsilon \tilde{I} = \tilde{\rho}_{se}\delta(z - z_0) \tag{2.20}$$

Equation (2.20) clearly indicates that the transmission line voltage is discontinuous at $z = z_0$. This is possible only if the transmission line is excited by a lumped voltage source connected in series at $z = z_0$, as shown in Figure 2.2. The magnitude of the voltage source, V_0, should be equal to $\tilde{\rho}_{se}$.

The problem is now simplified considerably. All the parameters of the transmission line and the magnitude of the voltage source are known. The line voltage at any point of the transmission line can be determined in a straightforward manner.

To find the line voltage inside a given layer, the source layer voltage is to be found first. Let y_{in}^+ and y_{in}^- are the input admittances at the two sides of the voltage source. Then the line voltage at $z = z_0^+$ will be

$$V(z = z_0^+) = \tilde{\rho}_{se}\,\frac{y_{\text{in}}^-}{y_{\text{in}}^m} \tag{2.21}$$

where $y_{\text{in}}^m = y_{\text{in}}^+ + y_{\text{in}}^-$ (the superscript m stands for the TM$_z$ mode).

Suppose $y_{\text{in}}^{(1)}$ is the input admittance at the interface of layer 1 and layer 2. The reflection coefficient Γ_1 at $z = z_1$ is then

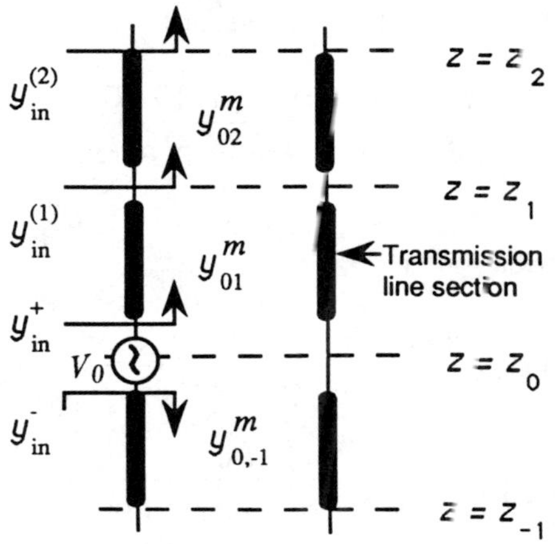

Figure 2.2 Spectral domain equivalent circuit of a layered structure for the TM_z mode with planar electric current excitation. The source plane is at $z = z_0$ and the source vector lies in the plane.

$$\Gamma_1 = \frac{y_{01}^m - y_{in}^{(1)}}{y_{01}^m + y_{in}^{(1)}} \tag{2.22}$$

The voltage distribution in the region $z_0 < z < z_1$, is given by

$$V_1(z) = V(z_0^+)\frac{[\exp\{-j(z - z_1)k_{1z}\} + \Gamma_1 \exp\{j(z - z_1)k_{1z}\}]}{[\exp\{-j(z_0 - z_1)k_{1z}\} + \Gamma_1 \exp\{j(z_0 - z_1)k_{1z}\}]} \tag{2.23}$$

where $V_1(z) = \tilde{D}_z(k_x, k_y, z)$ in the region $z_0 < z < z_1$. From (2.23) one can determine the line voltage at $z = z_1$. Repeating the procedure described previously, the line voltage inside layer 2 is derived as

$$V_2(z) = V_1(z_1)\frac{[\exp\{-j(z - z_2)k_{2z}\} + \Gamma_2 \exp\{j(z - z_2)k_{2z}\}]}{[\exp\{-j(z_1 - z_2)k_{2z}\} + \Gamma_2 \exp\{j(z_1 - z_2)k_{2z}\}]} \tag{2.24}$$

where $V_1(z_1)$ is the voltage at $z = z_1$, which can be obtained by setting $z = z_1$ in (2.23):

$$V_1(z_1) = \frac{V(z_0^+)(1 + \Gamma_1)}{[\exp\{jd_1k_{1z}\} + \Gamma_1 \exp\{-jd_1k_{1z}\}]}$$

where $d_1 = z_1 - z_0 =$ width of first layer.

Therefore, inside layer 2 the line voltage distribution is

$$V_2(z) = \frac{V(z_0^+)(1 + \Gamma_1)[\exp\{-j(z - z_2)k_{2z}\} + \Gamma_2 \exp\{j(z - z_2)k_{2z}\}]}{[\exp\{jd_1k_{1z}\} + \Gamma_1 \exp\{-jd_1k_{1z}\}][\exp\{jd_2k_{2z}\} + \Gamma_2 \exp\{-jd_2k_{2z}\}]}$$

where $d_2 = z_2 - z_1$ is the width of layer 2. For the ith layer, the generalized expression for the voltage distribution is

$$V_i(z) = V(z_0^+) \frac{\exp\{-j(z - z_i)k_{iz}\} + \Gamma_i \exp\{j(z - z_i)k_{iz}\}}{[\exp\{jd_ik_{iz}\} + \Gamma_i \exp\{-jd_ik_{iz}\}]} \cdot \prod_{n=1}^{i-1} \frac{(1 + \Gamma_n)}{[\exp\{jd_nk_{nz}\} + \Gamma_n \exp\{-jd_nk_{nz}\}]} \tag{2.25}$$

In the preceding expression $\Gamma_n = (y_{0n}^m - y_{\text{in}}^{(n)})/(y_{0n}^m + y_{\text{in}}^{(n)})$, $d_n = z_n - z_{n-1}$, and k_{nz} is the propagation constant of the transmission line associated with the n-th layer.

Because $V_i\,(z)$ is the two-dimensional Fourier transform of D_z, the electric displacement vector inside the ith layer should be

$$D_z(x, y, z) = \int_{-\infty}^{\infty} \int_{-\infty}^{\infty} V_i(z) \exp(-jk_x x - jk_y y)\, dk_x\, dk_y \tag{2.26}$$

where $V_i(z)$ is given in (2.25).

2.2.4 Construction of Complete Field Components

In Section 1.3, we showed that the TM_z field can be constructed from the z component of the electric displacement vector, D_z. The field components, in terms of A_z, are expressed in (1.11), where A_z and D_z are related via the following differential operator:

$$A_z = j\omega\mu \left(\frac{\partial^2}{\partial z^2} + k^2\right)^{-1} D_z \tag{2.27}$$

Taking the two-dimensional Fourier transformation of (2.27), one obtains

$$\tilde{D}_z = \frac{1}{j\omega\mu} \left(\frac{\partial^2}{\partial z^2} + k^2\right) \tilde{A}_z \tag{2.28}$$

Now, using (2.25), one can easily verify that inside the ith layer $\tilde{D}_z$ satisfies the following differential equation:

$$\frac{\partial^2 \tilde{D}_z}{\partial z^2} + k_{iz}^2 \tilde{D}_z = 0 \tag{2.29}$$

Equation (2.29) is a homogeneous, linear differential equation. Therefore, combining (2.28) and (2.29) one can show that $\tilde{A}_z$ also satisfies (2.29). That is,

$$\frac{\partial^2 \tilde{A}_z}{\partial z^2} + k_{iz}^2 \tilde{A}_z = 0 \tag{2.30}$$

Combining (2.28) and (2.30), one obtains

$$\tilde{A}_z = \frac{j\omega\mu\tilde{D}_z}{(k^2 - k_{iz}^2)}$$

In the space domain, A_z becomes

$$A_z = j\omega\mu \int_{-\infty}^{\infty}\int_{-\infty}^{\infty} \frac{\tilde{D}_z}{(k^2 - k_{iz}^2)} \exp(-jk_x x - jk_y y)\, dk_x\, dk_y$$

Invoking (1.11), one can write

$$B_x = \frac{\partial A_z}{\partial y} = \omega\mu \int_{-\infty}^{\infty}\int_{-\infty}^{\infty} \frac{k_y \tilde{D}_z}{(k^2 - k_{iz}^2)} \exp(-jk_x x - jk_y y)\, dk_x\, dk_y$$

$$B_y = \frac{-\partial A_z}{\partial x} = -\omega\mu \int_{-\infty}^{\infty}\int_{-\infty}^{\infty} \frac{k_x \tilde{D}_z}{(k^2 - k_{iz}^2)} \exp(-jk_x x - jk_y y)\, dk_x\, dk_y$$

$$B_z = 0$$

$$D_x = \frac{1}{j\omega\mu}\frac{\partial^2 A_z}{\partial x \partial z} = -j \int_{-\infty}^{\infty}\int_{-\infty}^{\infty} \frac{k_x}{(k^2 - k_{iz}^2)} \frac{\partial \tilde{D}_z}{\partial z} \exp(-jk_x x - jk_y y)\, dk_x\, dk_y$$

$$D_y = \frac{1}{j\omega\mu}\frac{\partial^2 A_z}{\partial y \partial z} = -j \int_{-\infty}^{\infty}\int_{-\infty}^{\infty} \frac{k_y}{(k^2 - k_{iz}^2)} \frac{\partial \tilde{D}_z}{\partial z} \exp(-jk_x x - jk_y y)\, dk_x\, dk_y$$

As a check, one can verify that inside a source free layer, $\nabla \cdot \mathbf{D} = 0$.

2.3 THE TE$_z$ FIELD PRODUCED BY PLANAR ELECTRIC CURRENT

In the preceding section, the TM$_z$ field components produced by a planar electric current source were derived. The complete field solution is the superposition of the TM$_z$ and TE$_z$ fields. This necessitates determination of the TE$_z$ field.

2.3.1 TE$_z$ Field Equation in the Fourier Domain

The TE$_z$ field solution can be constructed from the z component of the magnetic displacement vector, B_z. The partial differential equation satisfied by B_z is presented in Section 1.5 (see (1.29)). For a planar electric current, (1.29) reduces to

$$\nabla^2 B_z + k^2 B_z - \frac{\partial B_z}{\partial z}\frac{\partial \ln\mu}{\partial z} = -\mu\hat{z}\cdot(\nabla \times \mathbf{J}_e) \tag{2.31}$$

where $\mathbf{J}_e$ is the electric current density (amp/m^2). For a planar current source located at $z = z_0$, $\mathbf{J}_e$ can be represented as

$$\mathbf{J}_e = \mathbf{J}_{se}\delta(z - z_0) \tag{2.32}$$

where $\mathbf{J}_{se}$ represents the surface current density (amp/m). Let us now express B_z in terms of its two-dimensional Fourier transform $\tilde{B}_z$.

$$B_z(x, y, z) = \int_{-\infty}^{\infty}\int_{-\infty}^{\infty} \tilde{B}_z(k_x, k_y, z)\exp(-jk_x x - jk_y y)\,dk_x\,dk_y \tag{2.33}$$

where $\tilde{B}_z(k_z, k_y, z)$ is given by

$$\tilde{B}_z(k_x, k_y, z) = \frac{1}{4\pi^2}\int_{-\infty}^{\infty}\int_{-\infty}^{\infty} B_z(x, y, z)\exp(jk_x x + jk_y y)\,dx\,dy \tag{2.34}$$

Substitution of (2.33) in (2.31) yields

$$\int_{-\infty}^{\infty}\int_{-\infty}^{\infty}\left\{\frac{\partial^2\tilde{B}_z}{\partial z^2} + k_z^2\tilde{B}_z - \frac{\partial \ln\mu}{\partial z}\frac{\partial\tilde{B}_z}{\partial z}\right\}\exp(-jk_x x - jk_y y)\,dk_x\,dk_y = -\mu(\hat{z}\cdot\nabla \times \mathbf{J}_e) \tag{2.35}$$

where $k_z^2 = k^2 - k_x^2 - k_y^2$. Taking the inverse transform, one obtains the following equation for B_z:

$$\frac{\partial^2 \tilde{B}_z}{\partial z^2} + k_z^2 \tilde{B}_z - \frac{\partial \tilde{B}_z}{\partial z}\frac{\partial \ln\mu}{\partial z} = \frac{-\mu}{4\pi^2}\int_{-\infty}^{\infty}\int_{-\infty}^{\infty} \hat{z}\cdot(\nabla \times \mathbf{J}_e)\exp(jk_x x + jk_y y)\,dx\,dy \quad (2.36)$$

2.3.2 The Equivalent Transmission Line

Equation (2.36) and the magnetic field divergence equation are combined together to yield the telegrapher's equations pair. To that end, we express the divergence equation ($\nabla \cdot \mathbf{B} = 0$) in the Fourier domain, which is given by

$$\frac{\partial \tilde{B}_z}{\partial z} + \frac{\mu}{4\pi^2}\int_{-\infty}^{\infty}\int_{-\infty}^{\infty} (\nabla_t \cdot \mathbf{H}_t)\exp(jk_x x + jk_y y)\,dx\,dy = 0 \quad (2.37)$$

where $\mathbf{H}_t$ is the magnetic field vector with the longitudinal component suppressed; that is, $\mathbf{H}_t = \mathbf{H} - \hat{z}H_z \cdot \nabla_t$ is the transverse gradient operator.

$$\nabla_t = \hat{x}\frac{\partial}{\partial x} + \hat{y}\frac{\partial}{\partial y}$$

Denote the Fourier transform of $\nabla_t \cdot \mathbf{H}_t$ by $\tilde{T}$. We then have

$$\frac{\partial \tilde{B}_z}{\partial z} + \mu\tilde{T} = 0 \quad (2.38)$$

with

$$\tilde{T} = \frac{1}{4\pi^2}\int_{-\infty}^{\infty}\int_{-\infty}^{\infty} (\nabla_t \cdot \mathbf{H}_t)\exp(jk_x x + jk_y y)\,dx\,dy \quad (2.39)$$

Differentiating (2.38) with respect to z, we obtain

$$\frac{\partial^2 \tilde{B}_z}{\partial z^2} + \mu\frac{\partial \tilde{T}}{\partial z} + \tilde{T}\frac{\partial \mu}{\partial z} = 0 \quad (2.40)$$

Combining (2.36) and (2.40) we have

$$\frac{\partial \tilde{T}}{\partial z} - \frac{k_z^2}{\mu}\tilde{B}_z = \frac{1}{4\pi^2}\int_{-\infty}^{\infty}\int_{-\infty}^{\infty} \hat{z}\cdot(\nabla \times \mathbf{J}_e)\exp(jk_x x + jk_y y)\,dx\,dy$$

or

$$\frac{\partial \tilde{T}}{\partial z} - \frac{k_z^2}{\mu}\tilde{B}_z = I_0\delta(z - z_0) \tag{2.41}$$

where

$$I_0 = \frac{1}{4\pi^2}\int_{-\infty}^{\infty}\int_{-\infty}^{\infty} \hat{z}\cdot(\nabla \times \mathbf{J}_{se}) \exp(jk_x x + jk_y y)\, dx\, dy$$

is the two-dimensional Fourier transformation of $\hat{z}\cdot(\nabla \times \mathbf{J}_{se})$, $\mathbf{J}_{se}$ being the surface current density.

Equations (2.38) and (2.41) are telegrapher's equations pair. $\tilde{B}_z$ and $\tilde{T}$ can be regarded as the line voltage and the line current, respectively. In view of the preceding, the source is a shunt current of magnitude I_0, located at $z = z_0$. (Figure 2.3). The series impedance and the shunt admittance per unit length are μ and $-k_z^2/\mu$, respectively. The characteristic impedance of the equivalent transmission line is

$$y_0^e = \sqrt{-\frac{k_z^2}{\mu^2}} = \frac{jk_z}{\mu} \tag{2.42}$$

The propagation constant, β, is equal to k_z, when k_z is real. When k_z is imaginary, β should be considered as $-j|k_z|$ to ensure that the field decays in the positive z direction.[2]

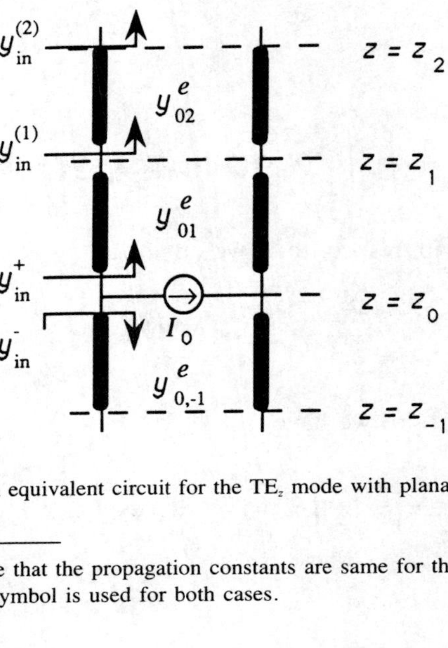

Figure 2.3 Spectral domain equivalent circuit for the TE$_z$ mode with planar electric current excitation.

[2]The reader may note that the propagation constants are same for the TM$_z$ and TE$_z$ transmission lines, and hence the same symbol is used for both cases.

Using Figure 2.3, one can determine the line voltage at any point of the equivalent transmission line. Recall that the transmission line voltage is equal to the Fourier transform of B_z, so that B_z is the Fourier inverse of the line voltage. In the following section, the transmission line voltage will be obtained in terms of I_0.

2.3.3 The Magnetic Displacement Vector Component, B_z

The z component of the magnetic displacement vector, B_z, can be found if its Fourier transform is known. The Fourier transform of B_z ($= \tilde{B}_z$) is the line voltage of the equivalent transmission line for the TE_z mode, shown in Figure 2.3. In this section, we will derive a general expression for $\tilde{B}_z$ as a function of z. In the space domain the quantity B_z can be obtained by taking the inverse transform of $\tilde{B}_z$.

The line voltage at any point of the transmission line in Figure 2.3 can be obtained using a procedure similar to the one described in Section 2.2.3. The line voltage in the region $z > z_0$ can be expressed in terms of the voltage $V(z_0^+)$ at $z = z_0^+$ (note, for the TE_z equivalent transmission line, $V(z_0^+) = V(z_0^-) = V(z_0)$, unlike the TM_z transmission line in Figure 2.2, where V is discontinuous at $z = z_0$). The voltage at $z = z_0$ is

$$V(z_0) = \frac{I_0}{y_{\text{in}}^e} \tag{2.43}$$

where $y_{\text{in}}^e = y_{\text{in}}^+ + y_{\text{in}}^-$, y_{in}^+ and y_{in}^- are the input admittances seen at the two sides of the source. The voltage distribution in the ith transmission line section will have an identical expression to (2.25). The reflection coefficient from the $(n + 1)$th layer to the nth layer in this case would be

$$\Gamma_n = \frac{y_{0n}^e - y_{\text{in}}^{(n)}}{y_{0n}^e + y_{\text{in}}^{(n)}} \tag{2.44}$$

where $y_{\text{in}}^{(n)}$ is the input admittance at the n and $n + 1$ junction, looking away from the source. The displacement vector component, B_z, is then

$$B_z(x, y, z) = \int_{-\infty}^{\infty}\int_{-\infty}^{\infty} V_i(z) \exp(-jk_x x - jk_y y)\, dk_x\, dk_y \tag{2.45}$$

2.3.4 Construction of the TE_z Field Components

Inside a given layer, the TE_z field components can be constructed from a vector function $\mathbf{F} = \hat{z}F_z$, where F_z is functionally related to B_z. The relation is presented in Section 1.3 (1.12). In the Fourier domain, the relation can be expressed as

$$\tilde{F}_z = \frac{j\omega\varepsilon\tilde{B}_z}{k^2 - k_z^2} \tag{2.46}$$

where $\tilde{F}_z$ represents the two-dimensional (x and y) Fourier transform of $F_z(x, y, z)$. In the Fourier domain, the field components are expressed in terms of $\tilde{F}_z$. In the Fourier domain, the differential operators $\partial/\partial x$ and $\partial/\partial y$ will be replaced by $-jk_x$ and $-jk_y$, respectively. Therefore, in the Fourier domain, (1.12) is equivalent to

$$\tilde{B}_x = \frac{-jk_x}{j\omega\varepsilon}\frac{\partial\tilde{F}_z}{\partial z} = \frac{-k_x}{\omega\varepsilon}\frac{\partial\tilde{F}_z}{\partial z}$$

$$\tilde{B}_y = \frac{-k_y}{\omega\varepsilon}\frac{\partial\tilde{F}_z}{\partial z}$$

$$\tilde{D}_x = jk_y\tilde{F}_z \qquad \tilde{D}_y = -jk_xF_z \qquad \tilde{D}_z = 0$$

In the space domain, the field components are obtained by taking the inverse Fourier transforms of these equations. Using (2.46) for $\tilde{F}_z$ and then taking Fourier inversion, one obtains

$$B_x(x, y, z) = -j\int_{-\infty}^{\infty}\int_{-\infty}^{\infty}\frac{k_x}{k^2 - k_z^2}\frac{\partial\tilde{B}_z}{\partial z}\exp(-jk_xx - jk_yy)\,dk_x\,dk_y$$

$$B_y(x, y, z) = -j\int_{-\infty}^{\infty}\int_{-\infty}^{\infty}\frac{k_y}{k^2 - k_z^2}\frac{\partial\tilde{B}_z}{\partial z}\exp(-jk_xx - jk_yy)\,dk_x\,dk_y$$

$$D_x(x, y, z) = -\omega\varepsilon\int_{-\infty}^{\infty}\int_{-\infty}^{\infty}\frac{k_y}{k^2 - k_z^2}\tilde{B}_z\exp(-jk_xx - jk_yy)\,dk_x\,dk_y \tag{2.47}$$

$$D_y(x, y, z) = \omega\varepsilon\int_{-\infty}^{\infty}\int_{-\infty}^{\infty}\frac{k_x}{k^2 - k_z^2}\tilde{B}_z\exp(-jk_xx - jk_yy)\,dk_x\,dk_y$$

$$D_z(x, y, z) = 0$$

2.4 THE TM$_z$ FIELD PRODUCED BY PLANAR MAGNETIC CURRENT

The magnetic current does not exist in reality because an isolated magnetic pole can never exist. However, for a class of radiation problem, the magnetic current concept is useful. With regard to the outside fields, an aperture is equivalent to electric and magnetic surface currents. The equivalent surface current concept[3] simplifies many aperture and related problems. The field solutions for electric surface current sources are considered in Sections 2.2 and 2.3. We now consider a magnetic surface current source. The TM_z field components produced by the source are derived in this section. For that, we should solve equation (1.18), after setting $\mathbf{J}_e = \rho_e = 0$. Equation (1.18) reduces to

$$\nabla^2 D_z + k^2 D_z - \frac{\partial D_z}{\partial z}\frac{\partial \ln\varepsilon}{\partial z} = \varepsilon\hat{z}\cdot(\nabla \times \mathbf{J}_m) \tag{2.48}$$

Note that (2.48) is the dual of (2.31) with the following replacements: $B_z \Rightarrow D_z$; $\mu \Rightarrow \varepsilon$; $J_e \Rightarrow -J_m$. Therefore, the equivalent transmission line for the TM_z mode excited by a planar magnetic current source will be similar to that of the TE_z mode excited by a planar electric current source. The equivalent transmission line circuit will be same as shown in Figure 2.3. The characteristic admittance for the ith transmission line section in this case would be y_{0i}^m, where $y_{0i}^m = jk_{iz}/\varepsilon_i$. The source current is

$$I_0 = -\frac{1}{4\pi^2}\iint \hat{z}\cdot(\nabla \times \mathbf{J}_{sm})\exp(jk_x x + jk_y y)\,dx\,dy \tag{2.49}$$

$\mathbf{J}_{sm}$ is the magnetic surface current density, which is given by $\mathbf{J}_m = \delta(z - z_0)\mathbf{J}_{sm}$.

In the present situation, $\tilde{D}_z$ is the line voltage. The complete field solution can be expressed in terms of $\tilde{D}_z(k_x, k_y, z)$, as shown in Section 2.2.4. Also, $\tilde{D}_z(k_x, k_y, z)$ can be expressed in terms of $\tilde{D}_z(k_x, k_y, z_0)$ using (2.25). $\tilde{D}_z(k_x, k_y, z_0)$ is given by $\tilde{D}_z(k_x, k_y, z_0) = I_0/y_{in}^m$, where $y_{in}^m = y_{in}^+ + y_{in}^-$. Here, y_{in}^+ and y_{in}^- are the input admittance at the two sides of the source current in Figure 2.3.

2.5 THE TE_z FIELD PRODUCED BY PLANAR MAGNETIC CURRENT

The magnetic current source we consider here lies in a plane perpendicular to the z-axis. Also, the current vector is perpendicular to the z-axis, so that $\mathbf{J}_m \cdot \hat{z} = 0$. Further, the electric current source is not present, so that we should substitute $\mathbf{J}_e = \rho_e = 0$ in (1.29). The differential equation then reduces to

$$\nabla^2 B_z + k^2 B_z - \frac{\partial B_z}{\partial z}\frac{\partial \ln\mu}{\partial z} = \frac{\partial \rho_m}{\partial z} - \rho_m\frac{\partial \ln\mu}{\partial z} \tag{2.50}$$

---

[3] R. F. Harrington, *Time-Harmonic Electromagnetic Fields*, McGraw-Hill, New York, 1961, Chapter 3, p. 106.

Again (2.50) is the dual of (2.1) with D_z, ε, and ρ_e as B_z, μ, and ρ_m, respectively. The transmission line equivalent circuit is similar to that shown in Figure 2.2. The characteristic admittance of the ith transmission line section in this case would be y_{0i}^e. The equivalent source voltage, V_0, in this case is

$$V_0 = \tilde{B}_z \text{ (at } z = z_0) = \tilde{\rho}_{sm}$$

where the magnetic volume charge density, ρ_m, is related to ρ_{sm}. The relation is $\rho_m = \rho_{sm}\ \delta(z - z_0)$, where ρ_{sm} represents the surface charge density on the surface $z = z_0$. The general expression for $\tilde{B}_z$ at any point on the transmission line can be obtained using the transmission line theory as discussed in Section 2.2.3. The complete TE_z field components are then obtained using (2.47). The spectral domain equivalent circuits and related parameters for the TE_z and TM_z modes are given in Table 2.1. The variables in the table are $k_i = k_0\sqrt{\varepsilon_i\mu_i}$, $k_\rho = \sqrt{k_x^2 + k_y^2}$, $\mathbf{k}_\rho = \hat{x}k_x + \hat{y}k_y$. In the other transformed domain, the shunt currents will be equal to the spectra of $\nabla \times \mathbf{J}_{se} \cdot \hat{z}$ and $-\nabla \times \mathbf{J}_{sm} \cdot \hat{z}$, respectively, for electric and magnetic current excitations.

Table 2.1
Transmission Line Equivalents for Planar Source Excitations.

	Equivalent Voltage V (z)	*Equivalent Current I (z)*	*Characteristic Admittance (y_{0i})*	*Equivalent Source (V_0, I_0)*	*Equivalent Circuit*
Electric current excitation (TM_z mode)	$\tilde{D}_z$	$-j(\mathbf{k}_\rho \cdot \tilde{\mathbf{E}}_t)$	$j\beta_i/\varepsilon_i$	Series voltage $V_0 = \tilde{\rho}_{se}$	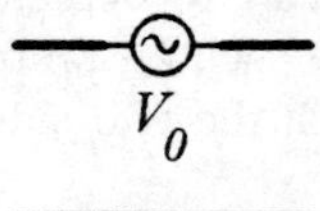
Electric current excitation (TE_z mode)	$\tilde{B}_z$	$-j(\mathbf{k}_\rho \cdot \tilde{\mathbf{H}}_t)$	$j\beta_i/\mu_i$	Shunt current $I_0 = -j(\mathbf{k}_\rho \times \tilde{\mathbf{J}}_{se}) \cdot \hat{z}$	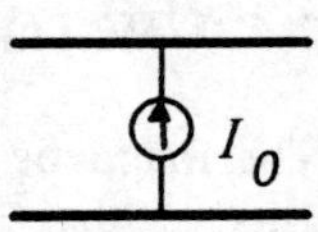
Magnetic current excitation (TM_z mode)	$\tilde{D}_z$	$-j(\mathbf{k}_\rho \cdot \tilde{\mathbf{E}}_t)$	$j\beta_i/\varepsilon_i$	Shunt current $I_0 = j(\mathbf{k}_\rho \times \tilde{\mathbf{J}}_{sm}) \cdot \hat{z}$	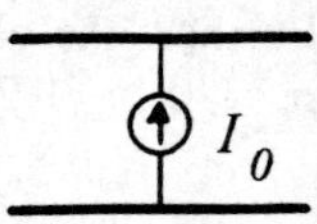
Magnetic current excitation (TE_z mode)	$\tilde{B}_z$	$-j(\mathbf{k}_\rho \cdot \tilde{\mathbf{H}}_t)$	$j\beta_i/\mu_i$	Series voltage $V_0 = \tilde{\rho}_{sm}$	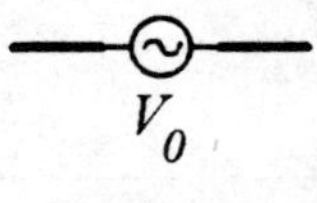

Note: The propagation constant is $\beta_i = \sqrt{k_i^2 - k_\rho^2}$.

2.6 A z-DIRECTED PLANAR CURRENT SOURCE

In Sections 2.2 and 2.3, we restricted ourselves to the current sources lying on a transverse plane ($z = z_0$ plane). The polarizations of the source current vectors were also in the transverse plane. In many cases, the current may not be in the transverse direction (for example, a monopole on a dielectric coated ground plane). Therefore we want to develop the procedure for obtaining fields produced by a z-directed current source. We will assume that the source length d is infinitely small[4] (Hertzian dipole) and the source lies on the $z = z_0$ plane. Mathematically, we can express the source as (here we consider the electric current source)

$$\mathbf{J} = \hat{z}I_z(x, y)\left\{u\left(z - z_0 + \frac{d}{2}\right) - u\left(z - z_0 - \frac{d}{2}\right)\right\}$$

where $I_z(x, y)$ is a function of x and y only. Here, $u(\cdot)$ represents the unit step function. For small d, the preceding current can equivalently be expressed in terms of current moment as $\mathbf{J} = \hat{z}(I_z d)\ \delta(z - z_0)$, where $I_z d$ is the current moment.

The TM_z field components can be obtained from D_z. The differential equation satisfied by D_z for this source will be

$$\nabla^2 D_z + k^2 D_z - \frac{\partial D_z}{\partial z}\frac{\partial \ln\varepsilon}{\partial z} = j\omega\mu\varepsilon I_z(x, y)\left\{u\left(z - z_0 + \frac{d}{2}\right) - u\left(z - z_0 - \frac{d}{2}\right)\right\} + \frac{\partial \rho_e}{\partial z} - \rho_e\frac{\partial \ln\varepsilon}{\partial z} \tag{2.51}$$

For the Hertzian dipole, two oppositely polarized electric charges will be accumulated at the two ends of the dipole source. The charge density ρ_e can be expressed as

$$\rho_e = \frac{-\nabla\cdot\mathbf{J}}{j\omega} = -\frac{1}{j\omega}\frac{\partial J_z}{\partial z} = -\frac{1}{j\omega}I_z(x, y)\left\{\delta\left(z - z_0 + \frac{d}{2}\right) - \delta\left(z - z_0 - \frac{d}{2}\right)\right\}$$

To solve for D_z in (2.51), we utilize the linearity property of linear differential equations. Let $D_z^{(1)}$ and $D_z^{(2)}$, respectively, be the solutions of the following two equations:

$$\nabla^2 D_z^{(1)} + k^2 D_z^{(1)} - \frac{\partial D_z^{(1)}}{\partial z}\frac{\partial \ln\varepsilon}{\partial z} = j\omega\mu\varepsilon I_z(x, y)\left\{u\left(z - z_0 + \frac{d}{2}\right) - u\left(z - z_0 - \frac{d}{2}\right)\right\} \tag{2.52}$$

[4]For an arbitrary source length, the superposition can be applied to find the total field produced by the source.

$$\nabla^2 D_z^{(2)} + k^2 D_z^{(2)} - \frac{\partial D_z^{(2)}}{\partial z}\frac{\partial \ln\varepsilon}{\partial z} = \frac{\partial \rho_e}{\partial z} - \rho_e \frac{\partial \ln\varepsilon}{\partial z} \tag{2.53}$$

Then, from linearity, one can write the complete solution for D_z as $D_z = D_z^{(1)} + D_z^{(2)}$.

Let us now try to find the solution of (2.52). We note that when d approaches 0, the righthand side of (2.52) reduces to

$$\nabla^2 D_z^{(1)} + k^2 D_z^{(1)} - \frac{\partial D_z^{(1)}}{\partial z}\frac{\partial \ln\varepsilon}{\partial z} = j\omega\mu\varepsilon d I_z(x, y)\delta(z - z_0) \tag{2.54}$$

Note that (2.54) and (2.48) have a common form. The spectral domain equivalent circuit corresponding to (2.54) will be the same as that shown in Figure 2.3. $D_z^{(1)}$ can be expressed as

$$D_z^{(1)} = \int_{-\infty}^{\infty}\int_{-\infty}^{\infty} \tilde{D}_z^{(1)} \exp(-jk_x x - jk_y y)\, dk_x\, dk_y$$

where $\tilde{D}_z^{(1)}$ is the transmission line voltage. The source current, I_0, in this case, would be

$$I_0 = \frac{-j\omega\mu d}{(2\pi)^2}\int_x\int_y I_z\,(x, y) \exp(jk_x x + jk_y y)\, dx\, dy \tag{2.55}$$

The line voltage, $\tilde{D}_z^{(1)}$ can be obtained from the equivalent circuit in Figure 2.3 with y_{0i}^e replaced by y_{0i}^m.

Toward the solution of (2.53), we note that (2.53) and (2.1) have the same form. In (2.1), the expression for ρ_e contains a single Dirac-delta function, whereas in (2.53) it contains two Dirac-delta functions. Therefore the equivalent circuit in the spectral domain will consist of two lumped voltage sources connected in series, as shown in Figure 2.4. The distance between two voltage sources should be d. The magnitude of V_0 is (as per (2.16)):

$$V_0 = \frac{1}{(2\pi)^2 j\omega}\int\int I_z(x, y) \exp(jk_x x + jk_y y)\, dx\, dy = \frac{\tilde{I}_z(k_x, k_y)}{j\omega} \tag{2.56}$$

One may note that, for a finite current moment $I_z d$, V_0 approaches an infinitely large value as d approaches 0. However, two infinitely large voltage sources with opposite polarity will yield finite values of the line voltage and line current (see Problem 2.14). The solution for $D_z^{(2)}$ is given by

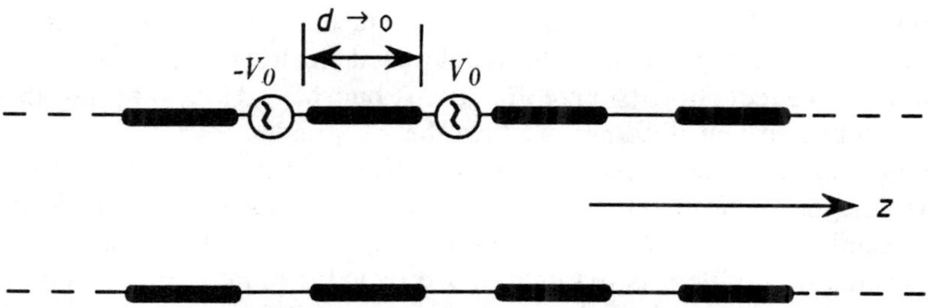

Figure 2.4 Equivalent circuit for solving $D_z^{(2)}$ from (2.53).

$$D_z^{(2)} = \int_{-\infty}^{\infty}\int_{-\infty}^{\infty} \tilde{D}_z^{(2)} \exp(-jk_x x - jk_y y)\, dk_x\, dk_y$$

where $\tilde{D}_z^{(2)}$ is the line voltage in the equivalent circuit in Figure 2.4. Because the circuit parameters are same in both cases, viz., for $\tilde{D}_z^{(1)}$ and $\tilde{D}_z^{(2)}$, we can use the superposition to obtain the complete equivalent circuit for D_z $(= \tilde{D}_z^{(1)} + \tilde{D}_z^{(2)})$. See Figure 2.5. The displacement vector component D_z is

$$D_z = \int_{-\infty}^{\infty}\int_{-\infty}^{\infty} \tilde{D}_z \exp(-jk_x x - jk_y y)\, dk_x\, dk_y$$

where $\tilde{D}_z$ is the line voltage in the circuit produced by the three sources shown.

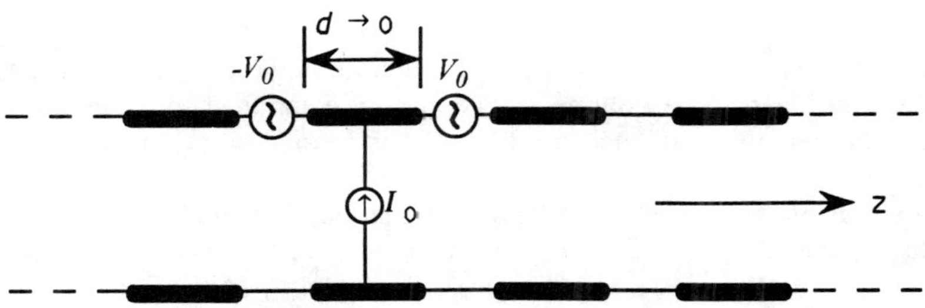

Figure 2.5 Complete spectral domain equivalent circuit of a layered structure excited by a planar electric current. The current vector is perpendicular to the z-plane.

In the preceding, we obtained the spectral domain equivalent circuit that yields $\tilde{D}_z$. From $\tilde{D}_z$ we can obtain the TM$_z$ mode fields in the usual manner. To obtain the

TE_z fields produced by the z-directed electric current element, we should solve for B_z. Recalling (1.29), we note that, for a z-directed electric current element, the right-hand side of the equation (source function) vanishes. This indicates that this z-directed electric current does not produce the TE_z mode.

Following a similar procedure, we can construct the equivalent circuit for the z-directed magnetic current element. This source produces only the TE_z mode. Therefore, we need to solve for B_z. Note from (1.29) and (1.18) that z-directed magnetic current excitation problem is the dual of z-directed electric current excitation problem. Therefore, the equivalent circuit, in this case, will be exactly same as in Figure 2.5. The current and voltage sources, however, will have different expressions, as compared to (2.55) and (2.56). Derivations of these expressions are left to the reader as an exercise.

2.7 PROCEDURE FOR VOLUME SOURCES

In the preceding sections, the procedure for obtaining fields excited by planar electric and magnetic surface currents were discussed. For volume sources, one can use the planar source model and then apply superposition to obtain the fields (as Maxwell's equation are linear, superposition is applicable). The volume source can be represented as a summation of planar sources, each of which lies on an infinitely thin layer. Let us illustrate this point in more detail. Suppose, we have a magnetic volume current source $\mathbf{M} = \hat{x}M_x(x, y, z)$. Using the property of the Dirac-delta function, this source can be expressed as

$$\mathbf{M} = \hat{x}\int_{z_1}^{z_2} M_x(x, y, z_0)\,\delta(z - z_0)\,dz_0 = \lim_{h\to 0}\hat{x}\sum_{n=1}^{N} M_x(x, y, z_0)\,h\delta(z - z_0)$$

with $z_0 = z_1 + nh$, and $N = [(z_2 - z_1)/h]$, where h is the thickness of the fictitious, infinitely thin layers. The volume source is now equivalent to a number of planar sources. The magnitude of the source, which is located at $z = z_0$, is $\mathbf{M}_{xs} = \hat{x}M_x(x, y, z_0)h$.

Now, using the procedure discussed in Sections 2.4 and 2.5, we can determine the fields produced by M_{xs}. Let $h\mathbf{E}_s$ and $h\mathbf{H}_s$ be the fields produced by M_{xs} Note that $\mathbf{E}_s(x, y, z|z_0)$ and $\mathbf{H}_s(x, y, z|z_0)$ are the fields produced by $M_x(x, y, z_0)$. Then the total field will be

$$\mathbf{E} = \lim_{h\to 0}\sum_{n=1}^{N}\mathbf{E}_s h \qquad \mathbf{H} = \lim_{h\to 0}\sum_{n=1}^{N}\mathbf{H}_s h$$

Under the limiting condition that h is approaches 0, $\mathbf{E}$ and $\mathbf{H}$ become

$$\mathbf{E} = \int_{z_1}^{z_2} \mathbf{E}_s(x, y, z|z_0)\, dz_0 \qquad \mathbf{H} = \int_{z_1}^{z_2} \mathbf{H}_s(x, y, z|z_0)\, dz_0$$

where $\mathbf{E}_s$ and $\mathbf{H}_s$ are the fields produced by a magnetic surface current source represented by $\mathbf{J}_m = \hat{x}M_x(x, y, z_0)\,\delta(z - z_0)$.

2.8 SUMMARY

In this chapter, methods for obtaining fields in layered media were presented. Equivalent transmission line circuits associated with TE_z and TM_z mode fields for planar electric and magnetic current excitations were obtained.[5] Procedures for nonplanar (volume) current sources were also outlined.

The analysis was carried out in the Fourier domain. However, the theory is also valid for other domains (such as Hankel domain[6]). For some special source configurations, it is convenient to express the fields in some particular domain than an other. As an example, for a circular source, the Hankel domain is more appropriate than the Fourier domain, though, in principle, both are applicable.[7] In the next chapter we will present an example to illustrate this point.

PROBLEMS

The reader is encouraged to solve these problems applying the spectral analysis, presented in this chapter, even though some problems can be solved using other methods.

2.1 A uniform, x-directed, planar electric surface current source is placed on the $z = 0$ plane. Obtain the TE_z and TM_z field components produced by the source. The medium is homogeneous in all directions with relative permittivity ε_r. Also, find the power per unit area that propagates along the positive z direction. (Hints: Use the equivalent circuits shown in Figures 2.2 and 2.3 for TM_z and TE_z fields, then calculate the Poynting vector for power per unit area.)

2.2 Recall Problem 2.1, consider that the current is placed above an electric conductor (ground plane). The distance between the ground plane and the source

[5]An alternate spectral domain transmission line method is also available in the literature. See T. Itoh and W. Menzel, "A Full Wave Analysis Method for Open Microstrip Structures," *IEEE Trans. on Antennas and Propagation*, Vol. AP-29, Jan. 1981, pp. 63–68.

[6]See J. Mathews and R.L. Walker, *Mathematical Methods of Physics*, 2d ed., W.A. Benjamin, Menlo Park, CA, 1979, Chapter 4.

[7]In fact, for circular geometry, the two dimensional Fourier transform simplifies to Hankel transform.

is d. Find the power per unit area that flows along z direction. Show that if $d = \lambda/4$, where λ is the wavelength inside the medium, the power per unit area will be 3 dB larger than that in Problem 2.1.

2.3 A uniform y-directed magnetic surface current source is place on the $z = 0$ plane. Show that only the TM_z mode fields will be produced. Draw the spectral domain equivalent circuit for the TM_z mode. Obtain the complete field components produced by this source.

2.4 A uniform x-directed electric surface current and a uniform y-directed magnetic surface current are placed on the $z = 0$ plane. Using superposition, obtain the total electric field in the regions $z > 0$ and $z < 0$. Show that by adjusting the amplitude of the two sources, it is possible to obtain a null field in the region $z < 0$. What would be the amplitude ratio of the sources to obtain a null field in the region $z < 0$? What are the corresponding electric and magnetic field components in the region $z > 0$?

2.5 A uniform x-directed magnetic surface current is placed at the interface of two media. The interface lies on the $z = 0$ plane. The dielectric constants of the two media are ε_1 and ε_2. Obtain the electric and magnetic field components in the two regions ($z > 0$ and $z < 0$) produced by the source.

2.6 Recall Problem 2.5, if an electric conductor plane is located at $z = -h$, what would be the field components in the regions $-h < z < 0$ and $z > 0$?

2.7 An x-directed uniform electric line current is placed in a vacuum. Show that the current will produce only the TE_z fields. Determine the electric field components. Show that E_x can be expressed as

$$E_x = -\frac{\omega\mu I_0}{4} H_0^{(2)}(k_0\rho)$$

where $\rho = \sqrt{y^2 + z^2}$ and I_0 is the magnitude of the line current. $H_0^{(2)}(k_0\,\rho)$ is the Hankel function of second kind of order 0.

2.8 An x-directed electric line current is placed at the air-dielectric interface. A conducting ground plane (electric conductor) is located at a distance d from the interface (on the dielectric side). Obtain the electric and magnetic field components produced by the source. The input impedance per unit length is defined as $z_{in} = -E_x/I_0$, where I_0 is the total current and E_x is the x-component of the electric field at the current location. What would be the input resistance per unit length seen by the source? Show that input reactance is infinitely large in this case.

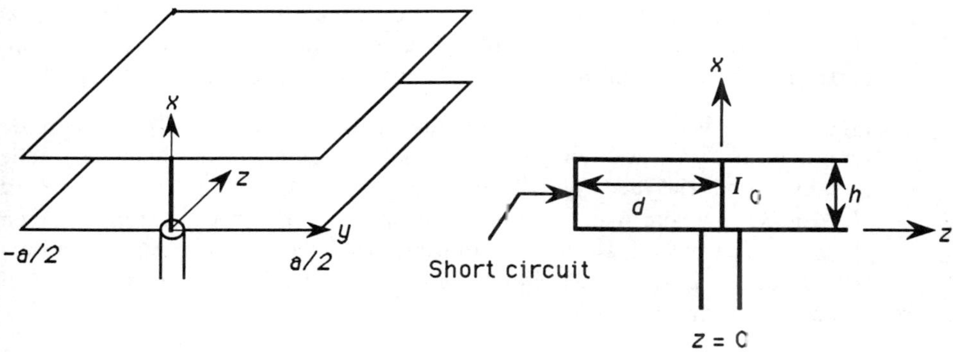

Figure 2.6 A parallel plate waveguide excited by a thin probe. The waveguide is shorted at $z = -d$.

2.9 The parallel plate waveguide in Figure 2.6 is excited by a very thin probe current (uniform along x). Assume magnetic walls at $y = \pm a/2$. Obtain the electric field components. Find the input resistance seen by the probe. (Hints: Expand the source current in Fourier series form in x-y plane. Obtain the fields for each harmonic and then use superposition. Note that to satisfy boundary conditions at for H_y, $y = \pm a/2$, I_x should be expressed as $I_x = \Sigma A_n \cos[n\pi(x - a/2)/a]$.)

2.10 The waveguide in Figure 2.7 is excited by a probe current. The current distribution on the probe is

$$I_y = \frac{I_0 \sin k_0 (d - y)}{\sin k_0 d}$$

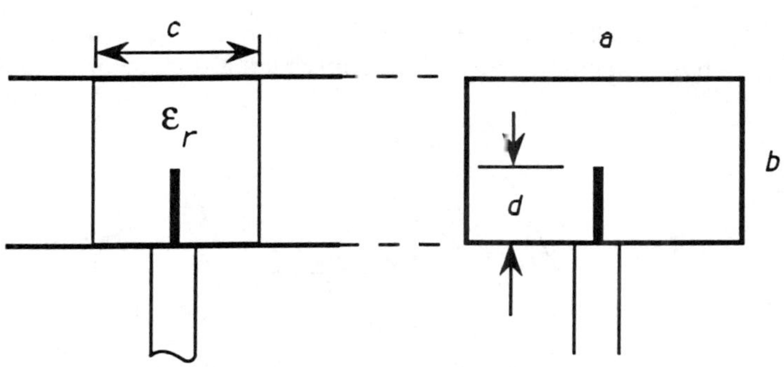

Figure 2.7 A rectangular waveguide excited by a thin probe. The probe is kept inside a dielectric slab and is parallel to the y-axis; $y = 0$ corresponds to the bottom wall of the waveguide.

where I_0 is the input probe current at $y = 0$. Obtain the dominant mode (TE_{10}) electric field inside and outside the dielectric region. If the waveguide supports only the dominant mode, find the input resistance seen by the probe.

2.11 Figure 2.8 shows a coaxial feed to circular waveguide adapter. The waveguide supports the TE_{z11} mode only. Obtain an expression for the input resistance seen by the coaxial feed. The distance between the probe and the short circuit plane is $\lambda/4$, where λ is the guide wavelength. The probe current is given by $I_y = -I_0 \sin(K_0 y)/\sin(K_0 d)$. (Hints: Expand the feed current in Fourier-Bessel series. Take the term that corresponds to the TE_{z11} mode and then proceed as in the previous problem.)

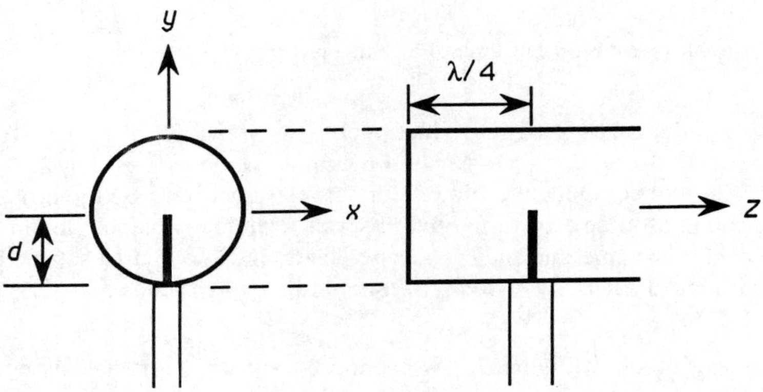

Figure 2.8 A coaxial feed to circular waveguide adapter. The waveguide is shorted at a distance of a quarter-guide wavelength from the transition.

2.12 A circular conductor, carrying a uniform current is placed over a dielectric slab as shown in Figure 2.9. The radius of the conductor is a. Obtain the

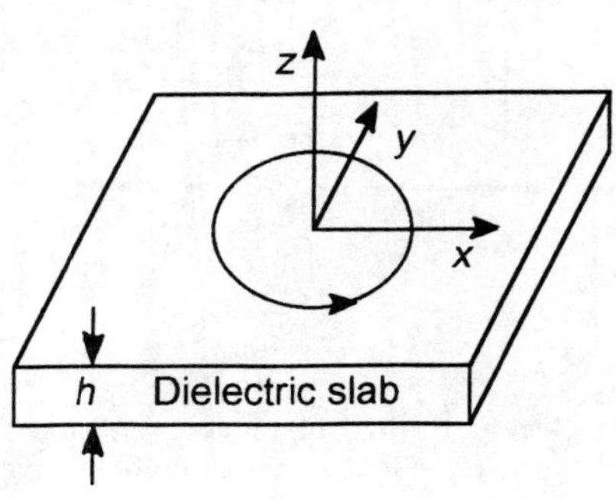

Figure 2.9 A circular electric current source over a dielectric slab.

electric and magnetic fields produced by this current in all three regions ($z > 0$, $-h < z < 0$, $z < -h$).

2.13 A z-directed electric current element (Hertzian dipole) is placed inside a dielectric slab as shown in Figure 2.10. Obtain the field components produced by the source. Assume that the length of the source is d, where d is small. Obtain D_z when d approaches 0.

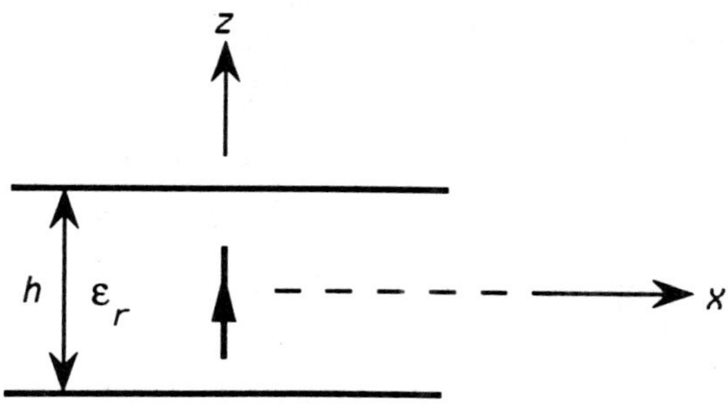

Figure 2.10 An electric current element inside a dielectric slab.

2.14 A z-directed Hertzian electric surface current source is placed inside a two-layered medium. The electrical parameters of the layers are (ε_1, μ_1) and (ε_2, μ_2), respectively. The source is located inside layer 2 at an infinitesimal distance from the interface. Show that in layer 2 the electric displacement vector component D_z is given by

$$D_z = \frac{-1}{\omega\varepsilon_2}\int_{-\infty}^{\infty}\int_{-\infty}^{\infty}\frac{(k_x^2 + k_y^2)\tilde{I}_m}{\left(\dfrac{k_{1z}}{\varepsilon_1} + \dfrac{k_{2z}}{\varepsilon_2}\right)}\exp(-jk_x x - jk_y y - jk_{2z} z)\, dk_x\, dk_y$$

where $\tilde{I}_m$ represents the Fourier transform of the current moment, and k_{1z} and k_{2z} are $\sqrt{k_0^2\mu_1\varepsilon_1 - k_x^2 - k_y^2}$ and $\sqrt{k_0^2\mu_2\varepsilon_2 - k_x^2 - k_y^2}$, respectively. Layer 1 exists in the region $z < 0$ and layer 2 exists in the region $z > 0$.

2.15 Obtain the spectral domain equivalent circuit for B_z where the source is a magnetic Hertzian dipole. The dipole is placed at the junction of the two layers. Find the solution for B_z.

Chapter 3

Current Sources over a Grounded Dielectric Structure

3.1 INTRODUCTION

This chapter illustrates applications of the spectral domain technique developed in Chapter 2. Two canonical problems are considered for this purpose. The first problem is a magnetic line current placed over a grounded planar dielectric structure. The expressions for the TM_z mode fields are obtained first. The theory of complex integration is then invoked to separate the space and surface wave fields. The radiated fields are obtained by applying the saddle point method of integration. The space and surface wave power produced by the source are then determined. The problem of uniform electric current source is solved in a similar manner. A procedure for circular sources is outlined at the end of the chapter.

3.2 MAGNETIC CURRENT OVER A GROUNDED DIELECTRIC STRUCTURE

In this section, we determine the fields produced by an infinitely long magnetic line current source located at the air-dielectric interface of a dielectric coated conducting ground plane as shown in Figure 3.1. The dielectric constant of the substrate is ε_r and h is the substrate thickness. Let $\mathbf{M} = M_0\hat{x}$ be the magnetic line current vector. Introducing the Dirac-delta function, the line current can be expressed as a surface current. The equivalent surface current density is given by

$$\mathbf{J}_{sm} = \hat{x}M_0\delta(y) \text{ V/m} \tag{3.1}$$

The equivalent volume current density is

$$\mathbf{J}_m = \mathbf{J}_{sm}\delta(z) = \hat{x}M_0\delta(y)\delta(z) \text{ V/m}^2 \tag{3.2}$$

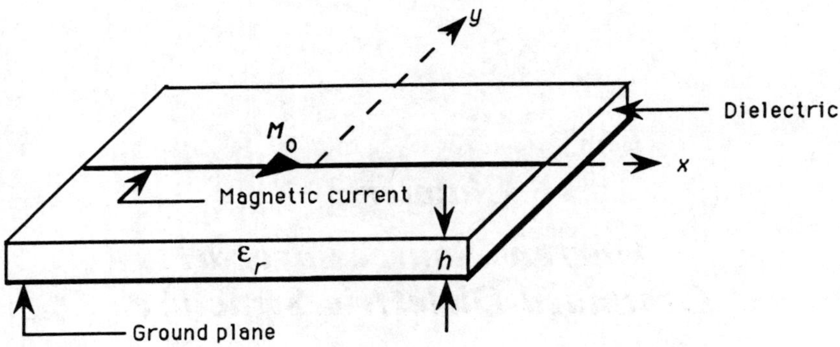

Figure 3.1 A magnetic line current over a dielectric coated ground plane.

First let us examine the type of fields produced by this source. Recall (1.19) in Chapter 1. Because $\hat{z} \cdot \nabla \times \mathbf{J}_m \neq 0$ in this case, the source will produce the TM_z mode field. Furthermore, we note that $\mathbf{J}_m$ satisfies (1.30), which implies that the source will not produce the TE_z mode field. Therefore we will consider the TM_z mode only.

3.2.1 The TM_z Mode Fields

The spectral domain equivalent circuit for the TM_z mode was developed in Section 2.4. Recall that in the Fourier domain D_z is considered the line voltage of the equivalent transmission line. Because $\tilde{D}_z$ is maximum on the ground plane of Figure 3.1, the ground plane is equivalent to an open circuit (see Section 1.4.1). The region $z > 0$ is equivalent to an infinitely long uniform transmission line. Figure 3.2 depicts the spectral domain equivalent circuit of the structure. The general expression for the source current I_0 in the equivalent circuit, given in (2.49), is

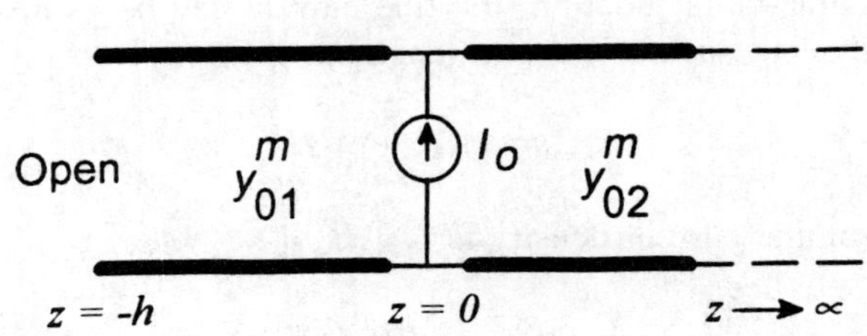

Figure 3.2 The spectral domain equivalent circuit for the TM_z mode fields.

$$I_0 = \frac{-1}{4\pi^2} \iint \hat{z} \cdot (\nabla \times \mathbf{J}_{sm}) \exp(jk_x x + jk_y y)\, dx\, dy$$

Using (3.1), the quantity $\hat{z} \cdot (\nabla \times \mathbf{J}_{sm})$ is obtained as

$$\hat{z} \cdot (\nabla \times \mathbf{J}_{sm}) = -M_0 \frac{\partial}{\partial y} \delta(y)$$

Substituting this expression for $\hat{z} \cdot (\nabla \times \mathbf{J}_{sm})$ in the integral and performing integration by parts, one obtains

$$I_0 = -\frac{1}{2\pi} jk_y M_0 \delta(k_x) \tag{3.3}$$

Note that we have used the following result in obtaining (3.3):

$$\int_{-\infty}^{\infty} \exp(jk_x x)\, dx = 2\pi\delta(k_x)$$

The next step is to find the voltage at $z = 0$. This can be done once the input admittance seen by I_0 is known. From Figure 3.2, the input admittance at $z = 0$ is obtained as $y_{\text{in}}^m = y_{02}^m + jy_{01}^m \tan(k_{1z}h)$. The voltage V_0 at $z = 0$ is $V_0 = I_0/y_{\text{in}}^m$.

The line voltage at any z can be expressed in terms of V_0. From Figure 3.2 one can visualize that, in the region $-h < z < 0$, the voltage distribution follows a standing wave pattern with an antinode (peak value) at $z = -h$. This is because the normal component of the electric field should have a maximum on a perfect electric conductor. The voltage distribution, therefore, is

$$V(z) = A \cos\{k_{1z}(z + h)\} \qquad -h < z < 0$$

Here, A is an unknown constant to be found. At $z = 0$, the voltage is V_0, which yields $A = V_0/\cos(k_{1z}h)$. Therefore, in the region $-h < z < 0$, the line voltage is given by

$$V(z) = \frac{V_0 \cos\{k_{1z}(z + h)\}}{\cos(k_{1z}h)}$$

Now consider the region $z > 0$. The voltage is purely progressive in nature due to the infinitely long transmission line. The voltage distribution is given by

$$V(z) = B \exp(-jk_{0z}z) \qquad z > 0$$

At $z = 0$, $V(z) = V_0$, which yields $B = V_0$. In general, the voltage distribution can be expressed as $V(z) = V_0 f(z)$, where

$$\begin{aligned} f(z) &= \frac{\cos\{k_{1z}(z + h)\}}{\cos(k_{1z}h)} \qquad -h < z < 0 \\ &= \exp(-jk_{0z}z) \qquad z > 0 \end{aligned} \tag{3.4}$$

From Table 2.1 recall that $\tilde{D}_z = V(z)$ for the TM_z mode, therefore, in the space domain D_z is given by

$$\begin{aligned} D_z(x, y, z) &= \int_{-\infty}^{\infty}\int_{-\infty}^{\infty} V_0 f(z) \exp(-jk_x x - jk_y y)\, dk_x\, dk_y \\ &= \int_{-\infty}^{\infty}\int_{-\infty}^{\infty} \frac{I_0}{y_{\text{in}}^m} f(z) \exp(-jk_x x - jk_y y)\, dk_x\, dk_y \end{aligned}$$

Substituting the expression for I_0 from (3.3) we obtain

$$D_z(x, y, z) = -\frac{jM_0}{2\pi}\int_{-\infty}^{\infty} \frac{k_y}{y_{\text{in}}^m} f(z) \exp(-jk_y y)\, dk_y \tag{3.5}$$

where

$$y_{\text{in}}^m = jk_{0z}/\varepsilon_0 - k_{1z}\tan(k_{1z}h)/\varepsilon_0\varepsilon_r \tag{3.6}$$

with $k_{0z}^2 = k_0^2 - k_y^2$ and $k_{1z}^2 = k_1^2 - k_y^2$.

Note that D_z is independent of x. This is expected, because the source current is uniform along x and extended to infinity in both directions. The TM_z field components are obtained from D_z using the procedure detailed in Chapter 1.

3.2.2 Singularities and Branch Points

The integrand in equation (3.5) has, in general, a number of singularities. The singularities are due to the zeros of the complex function $y_{\text{in}}^m(k_y)$. For a lossless medium, the singular points lie on the integration path (the real axis of the k_y plane) between k_0 and k_1. The number of positive zeros for y_{in}^m is[1] $N = \{n\} + 1$, where $\{n\}$ denotes the nearest integer smaller or equal to n, where n is given as

[1]The proof is left as an exercise for the reader.

$$n = \frac{2h}{\lambda_0}\sqrt{\varepsilon_r - 1}$$

and λ_0 is the wavelength in free space. Because y_{in}^m is a symmetric function of k_y, an equal number of negative zeros will also exist. The existence of zeros on the integration path makes the integral singular. To circumvent the problem of singularity, we consider that the dielectric substrate is lossy, having an infinitesimal loss tangent[2]. This makes k_0 complex with an infinitesimal negative imaginary part. For a complex k_0, the zeros of y_{in}^m lie off the real axis, and the integrand becomes regular on the real axis.

In addition to the singularity problem, k_{0z} is a double-valued function of k_y. The branch points are at $k_y = \pm k_0$. Because we have considered k_0 to be complex, the branch points also lie off the real axis. However, to compute the integral near a branch point, one needs to pay critical attention to the phase of k_{0z}, because its phase changes rapidly near that point.

3.2.3 Space and Surface Wave Fields

Equation (3.5) gives the total field produced by a magnetic line current on a grounded dielectric substrate. In this section we will show that the total field consists of two types of waves: the space wave and the surface wave. The *space wave* contributes to the far-field radiation. The *surface wave* propagates along the air-dielectric interface. The propagation behavior of a surface wave is similar to that of a guided wave inside a waveguide. In many situations we want to separate out these two waves. For example, suppose we want to examine the coupling between two sources located inside the same dielectric substrate. If the sources are not too close to one another, the coupling will be dominated by the surface wave fields. The contribution of the space wave will be insignificant. Therefore in this case we need to consider the surface wave fields only. On the other hand, if the second source lies much above the dielectric surface, the coupling will be dominated by the space wave.

To separate out the space and surface wave fields, the original path of integration in (3.5) is deformed. The deformed path (path *c*) is shown in Figure 3.3. Note that the branch point is kept outside the closed contour formed by the original path (real axis) and the deformed path. This makes the integrand "single valued" inside the closed contour, so that the theory of contour integration can be applied. The deformed path captures some singular points. From the contour integration theory, we write

[2]The loss tangent is defined as the ratio between the negative imaginary part and the real part of the complex dielectric constant. See R. E. Collin, *Field Theory of Guided Waves*, 2d ed., IEEE Press, New York, 1990, p. 198.

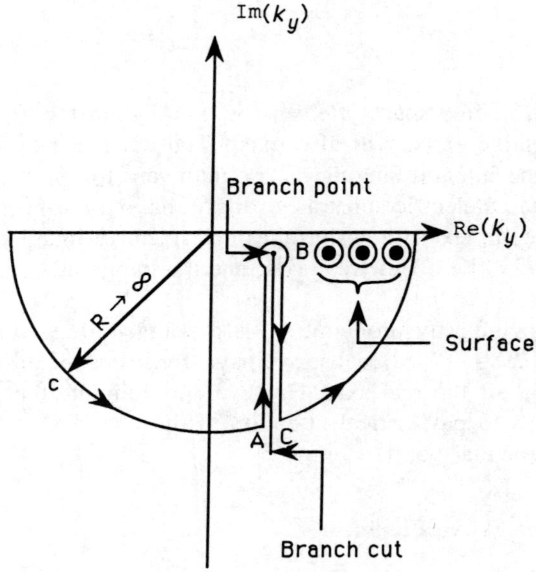

Figure 3.3 The deformed path of integration in the complex k_y plane. The deformed path facilitates separate out the space and surface wave fields.

$$\int_{-\infty}^{\infty} g(k_y)dk_y = \int_c g(k_y)dk_y - 2\pi j \sum_i \text{res}\{g(k_y), k_{yi}\} \tag{3.7}$$

where k_{yi} is the ith pole of $g(k_y)$ inside the closed region, and res$\{g(k_y), k_{yi}\}$ denotes the residue of $g(k_y)$ at $k_y = k_{yi}$. The residue is defined as

$$\text{res}\{g(k_y), k_{yi}\} = \lim_{k_y \to k_{yi}} [g(k_y)(k_y - k_{yi})] \tag{3.8}$$

Therefore, (3.5) can be expressed as

$$D_z = -\frac{jM_0}{2\pi} \int_c \frac{k_y}{y_{\text{in}}^m} f(z) \exp(-jk_y y)\, dk_y - M_0 \sum_{i=1}^{N} \text{res}\left[\frac{k_y}{y_{\text{in}}^m} f(z) \exp(-jk_y y), k_{yi}\right] \tag{3.9}$$

The first term on the righthand side of (3.9) is the space wave field, which has continuous spatial spectrum. Each term under the summation sign in (3.9) represents a surface wave mode. The number of surface wave modes is equal to the number of poles, N, inside the closed contour. Interestingly, the wave number along y is discrete in nature, similar to that of a guided wave. Also, the surface wave field decays exponentially along z. The surface wave field propagates along y and the propagation constant is k_{yi}.

3.2.4 Surface Wave Rays

The propagation of a surface wave mode can be viewed as the propagation of a plane wave[3] through multiple reflection as in a rectangular waveguide. Consider the ith surface wave mode. The field is given by

$$D_{zi}^{su} = -M_0 \operatorname{res}\left\{\frac{k_y f(z)\exp(-jk_y y)}{y_{in}^m}, k_{yi}\right\} \tag{3.10}$$

where $f(z)$ has two different forms in the two different regions, $z > 0$ and $z < 0$ (see (3.4)). Using L'Hospital's rule in (3.8), the surface wave field in the $z > 0$ region can be expressed as

$$D_{zi}^{su} = \frac{-M_0 k_{yi}\exp(-\gamma_{0zi}z - jk_{yi}y)}{dy_{in}^m/dk_y|_{k_y=k_{yi}}}$$

which, after simplification, becomes

$$D_{zi}^{su} = -M_0\varepsilon_0 T_{su}\exp(-\gamma_{0zi}z - jk_{yi}y) \tag{3.11}$$

where $\gamma_{0zi} = \sqrt{k_{yi}^2 - k_0^2}$, $k_{1zi} = \sqrt{k_1^2 - k_{yi}^2}$, and

$$T_{su} = \left[\frac{1}{\gamma_{0zi}} + \frac{\tan(k_{1zi}h)}{\varepsilon_r k_{1zi}} + \frac{h}{\varepsilon_r}\sec^2(k_{1zi}h)\right]^{-1}$$

The surface wave field in the region $z < 0$ will be

$$D_{zi}^{su} = -M_0\varepsilon_0 T_{su}\frac{\cos\{k_{1zi}(z+h)\}}{\cos(k_{1zi}h)}\exp(-jk_{yi}y) \tag{3.12}$$

When $z > 0$, the field decays exponentially. In the region $-h < z < 0$, the surface wave field has a standing wave pattern along z while propagating along y. This is

---

[3]In the present situation, where the source is uniform along x and extended to infinity, the equiphase surface of a surface wave field is planar. For a finite-dimensioned source, a surface wave field can be expressed as a cylindrical wave.

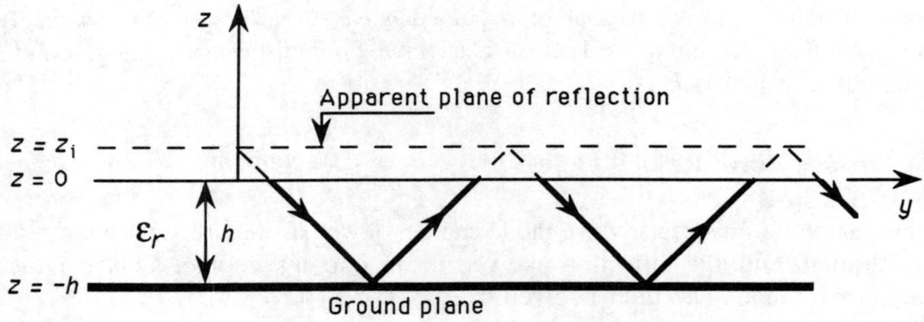

Figure 3.4 Propagation of surface wave fields through multiple reflections.

very similar to the propagation in a waveguide. This situation is equivalent to an interference of two obliquely incident plane wave fields. The angle of incidence, θ_i (see Figure 3.4), is given by[4] $\cos\theta_i = k_{yi}/k_1$.

The air-dielectric interface plane is not the apparent plane of reflection, because neither D_z^{su} nor its derivative (with respect to z) vanishes on this plane. The apparent plane of reflection can be considered as the plane on which the total field is 0 or maximum. This condition yields the location of the apparent plane of reflection for the ith surface wave mode. Suppose $z = z_i$ is the equation of the reflection plane. Then,

$$k_{1zi}(z_i + h) = \frac{n\pi}{2} \qquad n = 1, 2, 3, \ldots$$

or

$$z_i = \frac{n\pi}{2k_{1zi}} - h$$

The selection of n should be such that z_i becomes a positive number near 0, so that the reflecting surface is very close to the air-dielectric interface.

3.2.5 Radiated Fields

The far (radiated) field can be obtained by taking the asymptotic limit (y, $z \to \infty$) of the space wave field. From (3.9), the space wave field is expressed as

[4]Note that, because k_{yi} always less than k_1, θ_i is always real.

$$D_z^{sp} = \frac{-jM_0}{2\pi} \int_c \frac{k_y f(z) \exp(-jk_y y)}{y_{in}^m} dk_y \tag{3.13}$$

Because at present we are considering the field in the radiation zone, we have $z > 0$, so that $f(z) = \exp(-jk_{0z}z)$. Furthermore, for $y > 0$, $\exp(-jk_y y)$ approaches zero value on the circular segment of the contour shown in Figure 3.3. Therefore, D_z^{sp} is contributed by the segments around the branch cut only (ABC in Figure 3.3). Therefore, D_z^{sp} can be expressed in terms of the following complex line integral:

$$D_z^{sp} = \frac{-jM_0}{2\pi} \int_{ABC} \frac{k_y}{y_{in}^m} \exp(-jk_{0z}z - jk_y y) dk_y \tag{3.14}$$

The source produces the TM$_z$ fields only. Therefore, it is convenient to determine the magnetic vector potential A_z^{sp}, which yields complete field components for the TM$_z$ mode. Recalling Section 2.2.4, one can write

$$A_z^{sp} = \frac{M_0 \omega \mu}{2\pi} \int_{ABC} \frac{1}{k_y y_{in}^m} \exp(-jk_{0z}z - jk_y y) dk_y \tag{3.15}$$

because in this case $\tilde{A}_z = j\omega\mu\, (\tilde{D}_z / k_y^2)$.

We now use the following transformations of variables:

$$k_y = k_0 \sin\alpha \qquad y = \rho \sin\theta \qquad z = \rho \cos\theta$$

This yields

$$A_z^{sp} = \frac{M_0 \omega \mu \varepsilon_0}{2\pi k_0} \int_{c_1} \frac{\exp\{-jk_0\rho\cos(\theta - \alpha)\} \cot\alpha}{j\cos\alpha - \dfrac{\sqrt{\varepsilon_r - \sin^2\alpha}}{\varepsilon_r} \tan\{k_0 h \sqrt{\varepsilon_r - \sin^2\alpha}\}} d\alpha \tag{3.16}$$

where the contour c_1 is shown in Figure 3.5.

The path c_1 depends on the particular branch cut chosen. Note that a saddle point exists at $\alpha = \theta$. The saddle point is located near the contour c_1. Therefore, we deform the contour c_1 through the saddle point and use the saddle point[5] method of

[5]The deformed contour may capture complex poles that are associated with "leaky waves." Far-field radiation behavior of a source in a layered structure can be explained in terms of leaky wave excitations. See D. R. Jackson and A. A. Oliner, "A Leaky Wave Analysis of the High-Gain Printed Antenna Configuration," *IEEE Trans. on Antennas and Propagation*, Vol. AP-36, July 1988, pp. 905–910.

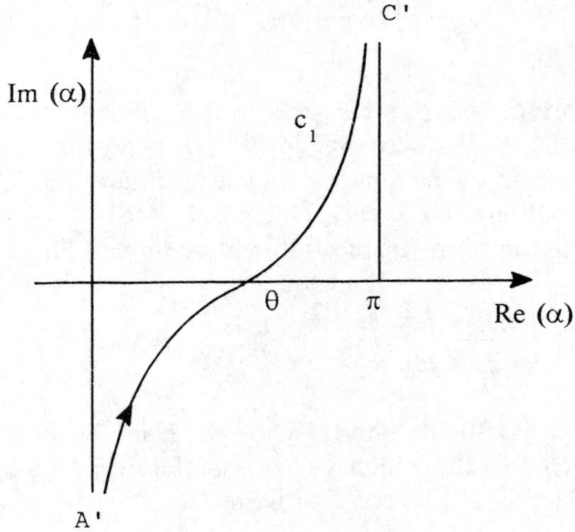

Figure 3.5 Mapping of the integration path from k_y plane to α plane, where $k_y = k_0 \sin \alpha$. Also shown is the path, c_1, through the saddle point.

integration to obtain the asymptotic expression for A_z^{sp}. The asymptotic expression for A_z^{sp}, considering ρ large, will be

$$A_z^{sp} \approx -j \frac{M_0 \mu}{2\eta \sin\theta} \exp(-jk_0\rho) \sqrt{\frac{2j}{\pi k_0 \rho}}\, F \tag{3.17a}$$

where the factor F is given by

$$F = \left[1 + j \frac{\sqrt{\varepsilon_r - \sin^2\theta}}{\varepsilon_r \cos\theta} \tan\left(k_0 h \sqrt{\varepsilon_r - \sin^2\theta}\right) \right]^{-1} \tag{3.17b}$$

Using the relation $\mathbf{H} = (\nabla \times \mathbf{A})/\mu$ and $\nabla \times \mathbf{H} = j\omega\varepsilon\mathbf{E}$ and ignoring the higher order terms, one obtains

$$E_\theta = \frac{k_0 M_0}{4} \sqrt{\frac{2j}{\pi k_0 \rho}} \exp(-jk_0\rho)\, 2F \tag{3.18}$$

Interestingly, the effect of the grounded dielectric layer on the far-field pattern appears through the factor $2F$. This factor may be called the *structure radiation factor* (SRF). When h is set to 0, the factor becomes equal to 2 and the radiated field becomes equal to that of a uniform magnetic line current placed on an infinite ground plane. If we look closely, factor F basically originates from the term $y^m_{\text{in}}(k_y)$, which solely depends on the layered structure and is independent of the source. Therefore, to determine the TM_z radiation field for an arbitrary planar source on a grounded dielectric layer, one can ignore the structure, determine the far-field first, and then multiply the SRF with the result.

As discussed earlier, the SRF yields the effect of the substrate and the ground plane on the radiation pattern. When $h > 0$ and $\varepsilon_r > 1$, the SRF may vary from 2 to 0 when θ varies from 0 to 90°. This means that the factor provides an additional taper to the radiation pattern. In Figure 3.6, we have plotted the magnitude of $2F$ versus the look angle θ. For a thin substrate, the curve is mostly flat, and the value of $|2F|$ decreases sharply in the close vicinity of $\theta = 90°$. For a thicker substrate, the flat region decreases and the value of $|2F|$ starts decreasing at a lower value of

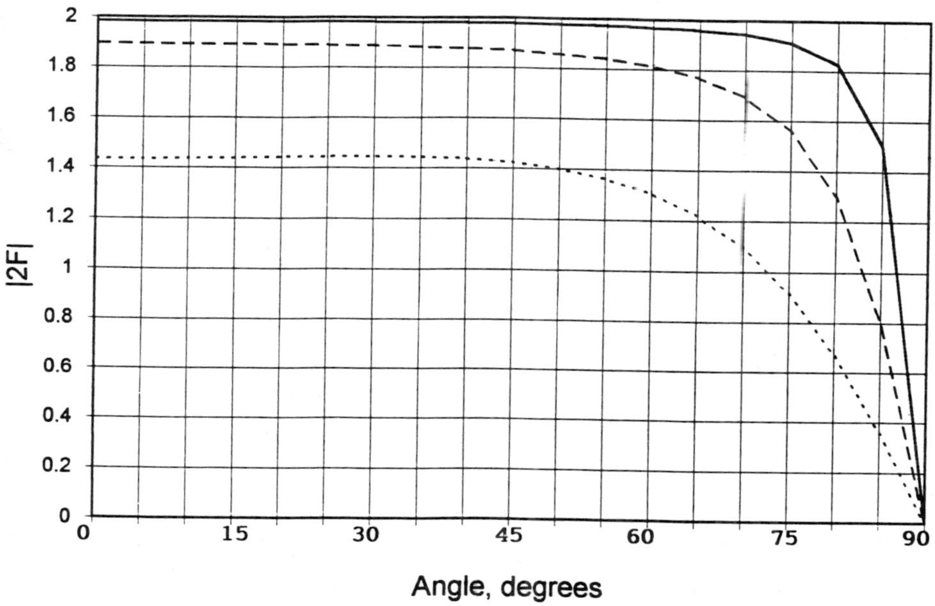

Figure 3.6 Variation of the structure radiation factor, $|2F|$, with the angle of radiation for the TM_z fields for $\varepsilon_r = 2.5$: —— $h = 0.02\lambda_0$, — — — $h = 0.05\lambda_0$, ---- $h = 0.1\lambda_0$.

θ. This results in a net increase in the directivity. The effect becomes more intensive if a superstrate is used in addition to a substrate. In fact, the directivity could be considerably higher than normal if an appropriate dielectric superstrate-substrate combination were used.[6]

3.2.6 Space and Surface Wave Power

The space wave power (radiated power) per unit length is calculated by integrating the Poynting vector with respect to θ. This calculation therefore would be

$$P^{\text{sp}} = \int_{-\pi/2}^{\pi/2} \frac{|E_\theta|^2}{\eta} \rho \, d\theta = 2 \int_0^{\pi/2} \frac{|E_\theta|^2}{\eta} \rho \, d\theta \quad \text{watts/m}$$

where $\eta = 120\pi\Omega$ is the free-space impedance. Invoking (3.18) for E_θ, one obtains

$$P^{\text{sp}} = \frac{k_0 |M_0|^2}{120\pi^2} \int_0^{\pi/2} |F|^2 \, d\theta \quad \text{watts/m}$$

where F is given in (3.17b).

To calculate the surface wave power, we obtain the surface wave field first. For the ith surface wave mode (TM_z type) the fields are derivable from D_{zi}^{sp}. Recalling (1.11), we write

$$B_{xi}^{\text{su}} = \frac{\partial A_{zi}^{\text{su}}}{\partial y} = \frac{j\omega\mu}{k_{yi}^2} \frac{\partial D_{zi}^{\text{su}}}{\partial y}$$

Using (3.11), we obtain

$$B_{xi}^{\text{su}} = \frac{\omega\mu}{k_{yi}} \{-M_0 \varepsilon_0 T_{\text{su}} \exp(-\gamma_{0zi} z - jk_{yi} y)\}$$

[6] D. R. Jackson and N. G. Alexopoulos, "Gain Enhancement Methods for Printed Circuit Antennas," *IEEE Trans. on Antennas and Propagation*, Vol. AP-33, No. 9, 1985, pp. 976–987. H. Y. Yang and N. G. Alexopoulos, "Gain Enhance Methods for Printed Circuit Antennas Through Multiple Superstrates," *IEEE Trans. on Antennas and Propagation*, Vol. AP-35, No. 7, 1987, pp. 860–863. A. K. Bhattacharyya, "Radiation Efficiency and the Gain of a Source Inside a Two-Layered Grounded Dielectric Structure," *Microwave and Optical Technology Letters*, Vol. 5, No. 9, Aug. 1992, pp. 439–441.

The surface wave power per unit length, for the ith mode is $P_i^{su} = -M_0 H_{xi}^{su*}$ ($y = z = 0$), which yields

$$P_i^{su} = \frac{k_0}{k_{yi}} \cdot \frac{|M_0|^2}{\eta} T_{su} \quad \text{watts/m}$$

In Figure 3.7 we have plotted the space and surface wave power versus the normalized substrate thickness. As expected, the surface wave power increases with the substrate thickness, making the radiator less efficient due to the surface wave loss.

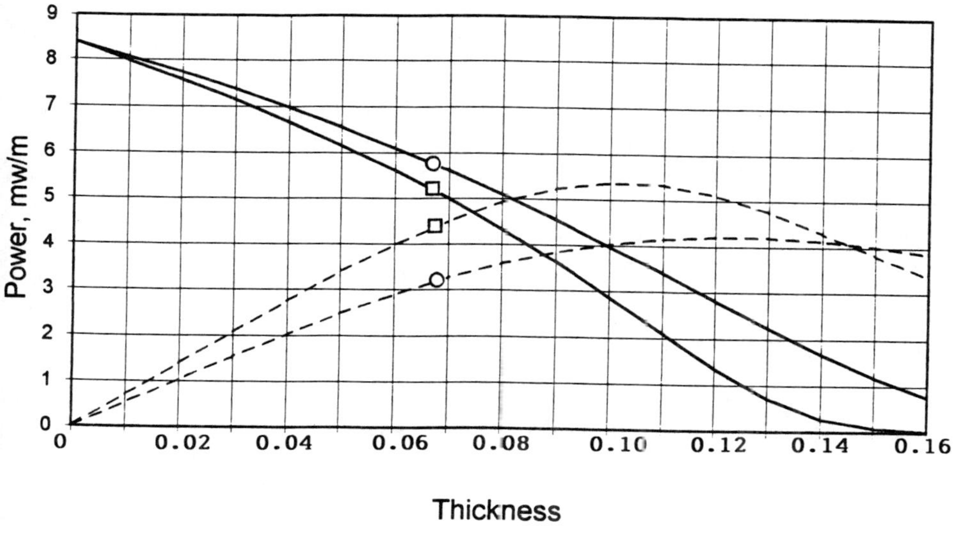

Figure 3.7 The TM_z mode space and surface wave power produced by a magnetic line current versus normalized substrate thickness, h/λ_0: $M_0 = 1$ volt, $\lambda_0 = 1$ meter; —○— space wave power ($\varepsilon_r = 2$), —□— space wave power ($\varepsilon_r = 3$), --○-- surface wave power ($\varepsilon_r = 2$), and --□-- surface wave power ($\varepsilon_r = 3$).

3.3 ELECTRIC CURRENT OVER A GROUNDED DIELECTRIC STRUCTURE

The electromagnetic fields produced by an electric line current are determined in this section. The current is assumed to be located at the air-dielectric interface of a grounded dielectric structure (as in Figure 3.1, except the magnetic current should be replaced by an electric current). This source produces the TE_z mode fields only. Therefore, the field components can be obtained from the z-directed magnetic displacement vector, B_z. The present problem is dual to that in Section 3.2. Using duality, we can write the expression for B_z from the expression of D_z in (3.5). With reference to Table 2.1 and (3.5), the expression for B_z would be

$$B_z = \frac{jI_0}{2\pi} \int_{-\infty}^{\infty} \frac{k_y}{y_{\text{in}}^e} g(z) \exp(-jk_y y)\, dk_y \tag{3.19}$$

where y_{in}^e in this case will be

$$y_{\text{in}}^e = \frac{jk_{0z}}{\mu_0} + \frac{k_{1z}}{\mu_0 \mu_r} \cot(k_{1z} h) \tag{3.20}$$

The sign difference between (3.5) and (3.19) is due to two different types of currents (magnetic and electric). Furthermore, the term y_{in}^e is different from y_{in}^m in (3.6). In the present case, $\tilde{B}_z$ is considered as the line voltage so that at $z = -h$, B_z should be 0, which is equivalent to a short circuit. In the spectral domain equivalent circuit, the input admittance seen by the source will be different. Also, $g(z)$ will have the following form due to the presence of the short circuit at $z = -h$:

$$\begin{aligned} g(z) &= \frac{\sin k_{1z}(z+h)}{\sin k_{1z} h} \qquad -h \le z \le 0 \\ &= \exp(-jk_{0z} z) \qquad z \ge 0 \end{aligned} \tag{3.21}$$

3.3.1 TE$_z$ Surface Wave Mode

To determine the TE$_z$ surface wave field, one has to obtain the zeros of y_{in}^e. It can be proven that at least one zero of y_{in}^e exists if the following condition holds:

$$\frac{4h\sqrt{\varepsilon_r - 1}}{\lambda_0} > 1$$

Interestingly, unlike the TM$_z$ surface wave mode, the TE$_z$ surface wave does not always exist in a grounded dielectric structure. The ith TE$_z$ surface wave mode is given by

$$B_{zi}^{\mathrm{su}} = I_0 \operatorname{res}\left\{\frac{k_y g(z)}{y_{\mathrm{in}}^e} \exp(-jk_y y),\ k_{yi}\right\}$$

$$= \mu_0 I_0 T'_{\mathrm{su}} \exp(-\gamma_{0zi} z - jk_{yi} y) \qquad z > 0 \tag{3.22}$$

$$= \mu_0 I_0 T'_{\mathrm{su}} \frac{\sin\{k_{1zi}(z+h)\}}{\sin(k_{1zi}h)} \exp(-jk_{yi} y) \qquad z < 0 \tag{3.23}$$

where k_{yi} is the ith zero of y_{in}^e. The other terms in (3.23) are $\gamma_{0zi} = \sqrt{(k_{yi}^2 - k_0^2)}$, $k_{1zi} = \sqrt{(k_1^2 - k_{yi}^2)}$, and

$$T'_{\mathrm{su}} = \left[\frac{1}{\gamma_{0zi}} - \frac{\cot(k_{1zi}h)}{\mu_r k_{1zi}} + \frac{h}{\mu_r} \operatorname{cosec}^2(k_{1zi}h)\right]^{-1}$$

The other TE_z surface wave field components can be derived from B_{zi}^{su}.

3.3.2 The TE_z Far Fields

The TE_z far field is obtained by evaluating the asymptotic limit of F_z, which is related to B_z via the relation

$$\tilde{F}_z = \frac{j\omega\varepsilon}{k_y^2} \tilde{B}_z$$

Following the procedure described for the TM_z far field, we obtain the far-field component

$$H_\theta = -\frac{k_0 I_0}{4} \sqrt{\frac{2j}{\pi k_0 \rho}} \exp(-jk_0\rho) \cdot 2F' \tag{3.24}$$

where

$$F' = \left[1 - j\frac{\sqrt{\varepsilon_r - \sin^2\theta}}{\mu_r \cos\theta} \cot(k_0 h \sqrt{\varepsilon_r - \sin^2\theta})\right]^{-1}$$

The structure radiation factor $2F'$ is due to the grounded dielectric layer. Note that when $\varepsilon_r = 1$, $2F'$ is equal to the space factor or array factor of two sources situated at a distance $2h$ from one another with a $180°$ phase difference. The second source is the negative image of the primary source created by the ground plane. Therefore, for the TE_z far field, the factor $2F'$ includes of the effect of the structure, which is independent of the source. In Figure 3.8, the magnitude of $2F'$ is plotted against the normalized substrate thickness. As can be noted, the grounded dielectric layer will alter the pattern considerably, especially for thick substrates.

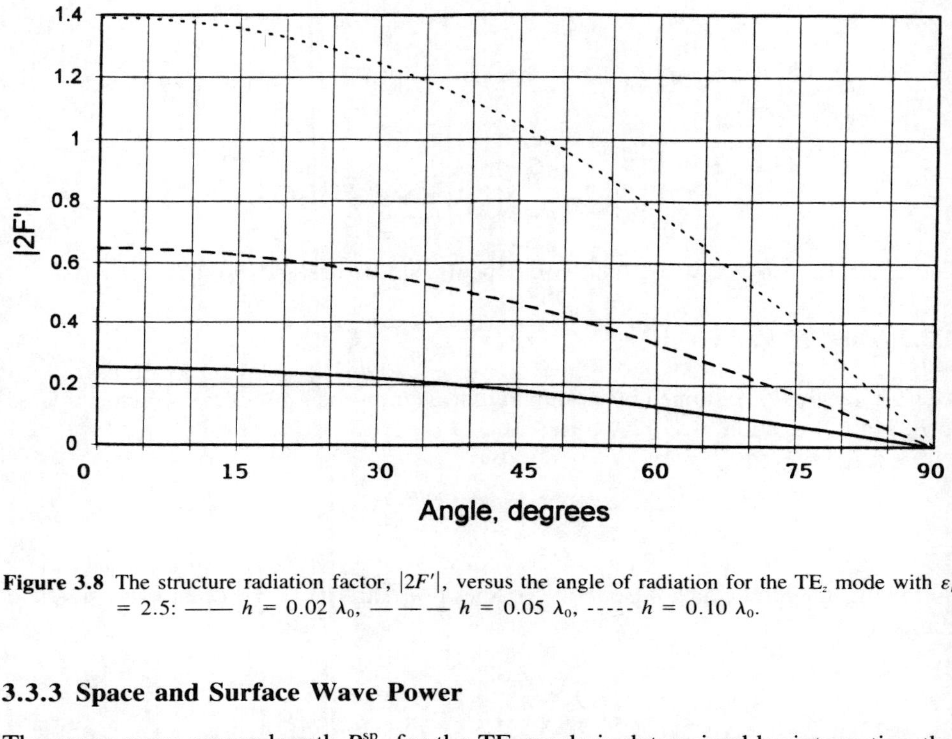

Figure 3.8 The structure radiation factor, $|2F'|$, versus the angle of radiation for the TE_z mode with $\varepsilon_r = 2.5$: —— $h = 0.02\ \lambda_0$, — — — $h = 0.05\ \lambda_0$, ----- $h = 0.10\ \lambda_0$.

3.3.3 Space and Surface Wave Power

The space wave power length P^{sp}, for the TE_z mode is determined by integrating the Poynting vector with respect to θ:

$$P^{sp} = \int_{-\pi/2}^{\pi/2} \eta |H_\theta|^2 \rho \, d\theta \tag{3.25}$$

Using (3.24), this expression is simplified to

$$P^{sp} = 120\, k_0 |I_0|^2 \int_0^{\pi/2} |F'|^2 \, d\theta \tag{3.26}$$

To determine ith mode surface wave power per unit length, we need to obtain E_{xi}^{su} at the source location. In view of (1.12)

$$D_x = -\frac{\partial F_z}{\partial y} = -\frac{j\omega\varepsilon}{k_y^2}\frac{\partial B_z}{\partial y}$$

we obtain

$$E_{xi}^{su} = -\frac{j\omega}{k_{yi}^2}\frac{\partial B_{zi}^{su}}{\partial y}$$

Therefore, from (3.22), we obtain

$$E_{xi}^{su}(y = z = 0) = -\mu_0 I_0 T'_{su}\frac{\omega}{k_{yi}}$$

The surface wave power, P_i^{su} will be

$$P_i^{su} = -E_{xi}^{su} I_0^* = \frac{k_0}{k_{yi}}|I_0|^2 T'_{su}\eta \tag{3.27}$$

In Figure 3.9 the space and surface wave power are plotted against the normalized substrate thickness with ε_r as a parameter. As noted, the surface wave power in-

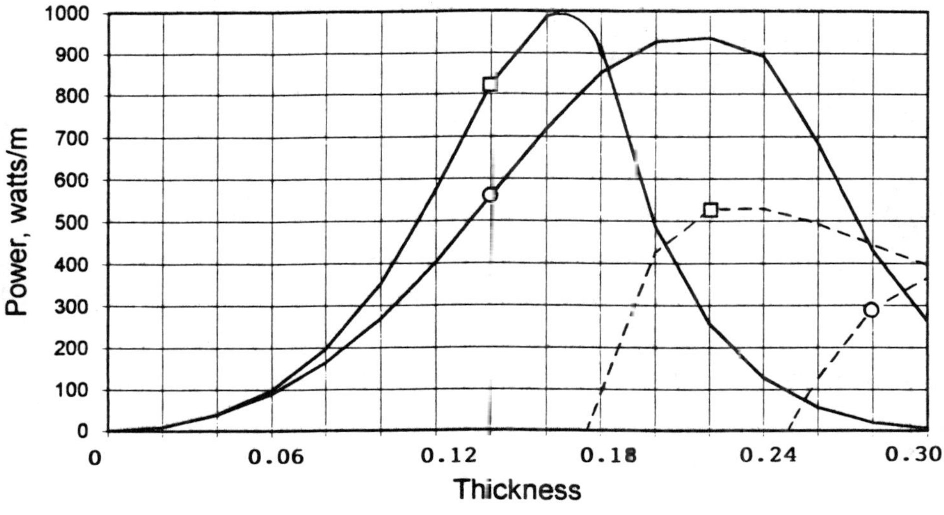

Figure 3.9 The TE$_z$ mode power produced by an electric line current versus normalized substrate thickness, h/λ_0: $I_0 = 1$ ampere, $\lambda_0 = 1$ meter; —○— space wave power ($\varepsilon_r = 2$), --□-- surface wave power ($\varepsilon_r = 3$), --○-- surface wave power ($\varepsilon_r = 2$), and —□— space wave power ($\varepsilon_r = 3$).

creases with ε_r or h. This implies that the radiation efficiency decreases if one of these quantities (ε_r or h) increases.

3.4 CIRCULAR SOURCES

In Sections 3.2 and 3.3, we considered straight line sources. We found that the Fourier transformed domain was very convenient to handle such problems. If the source contour (for line current source) or the source plane (for surface current source) has some other regularly shaped geometry (like a circle or ellipse), the Fourier transform of a function defined in the source region simplifies to some other transform, which appears to be more convenient to work with. For example, consider a function $f(x, y)$ defined in the circular region $\sqrt{(x^2 + y^2)} \leq a$. The Fourier transform of $f(x, y)$ would be

$$\tilde{f}(k_x, k_y) = \frac{1}{4\pi^2} \iint\limits_{\sqrt{(x^2 + y^2)} \leq a} f(x, y) \exp(jk_x x + jk_y y)\, dx\, dy \tag{3.28}$$

Without a loss of generality, we can assume the following form for $f(x, y)$.[7]

$$f(x, y) = g(\rho) \cos n\phi \tag{3.29}$$

where $\rho = \sqrt{(x^2 + y^2)}$, $\phi = \tan^{-1}(y/x)$. We now use the following transformations of variables:

$$k_x = k_\rho \cos \alpha \qquad k_y = k_\rho \sin \alpha \qquad x = \rho \cos \phi \qquad y = \rho \sin \phi$$

So that $dx\, dy$ is replaced by $\rho\, d\rho\, d\phi$. Then (3.28) becomes

$$\tilde{f}(k_\rho, \alpha) = \frac{1}{(2\pi)^2} \int_0^{2\pi} \int_0^a g(\rho) \cos n\phi \exp\{jk_\rho \rho \cos(\phi - \alpha)\}\, \rho\, d\rho\, d\phi \tag{3.30}$$

Now, the exponential term inside the integral can be expanded as

$$\exp\{jk_\rho \rho \cos(\phi - \alpha)\} = J_0(k_\rho \rho) + 2 \sum_{n=1}^{\infty} j^n J_n(k_\rho \rho) \cos n(\phi - \alpha) \tag{3.31}$$

[7]Any function can be expanded in Fourier-Bessel series. Here, we consider the terms with azimuthal variation as $\cos(n\phi)$.

where $J_n(\cdot)$ represents the nth order Bessel function of the first kind. Substituting (3.31) in (3.30) and integrating, we derive

$$\tilde{f}(k_\rho, \alpha) = \frac{j^n}{2\pi} \cos n\alpha \int_0^a g(\rho) J_n(k_\rho \rho) \rho \, d\rho \tag{3.32}$$

The integral in (3.32) is known as the nth order Hankel transform of the radial function $g(\rho)$. Note that the two-dimensional Fourier transform reduces to a one-dimensional Hankel transform. This is particularly true when the ϕ variation can be represented by a single cosine or sine term. In general, if all harmonics along ϕ are considered, the Fourier transformation will be equal to a summation of Hankel integrals, summed over the order of the harmonics. However, in many practical situations, the current distribution is dominated by only one cosine (or sine) variation along ϕ, so that only one term yields a fairly good approximation. This is especially true for circular patch antennas, where single-mode approximation provides fairly accurate results. Therefore, for circular geometry, it is convenient to use the Hankel domain directly rather than the Fourier domain. This will reduce the computational effort considerably. Because the Hankel transform evolved from the Fourier transform, all the formulations in Chapter 2 are equally valid in the Hankel domain. The Hankel domain analysis will be considered in further details in Chapter 5.

PROBLEMS

3.1 Consider a uniform magnetic line current source placed halfway between the ground plane and the air dielectric interface. The current is x- directed, as in Figure 3.1.

(a) Construct the spectral domain equivalent circuit for the TM_z mode.
(b) Obtain $y_{\text{in}}^m\ (k_y)$.
(c) Show that the surface wave poles do not depend on the location of the source.

3.2 Show that the surface wave poles (roots of $y_{\text{in}}^m(k_y) = 0$) lie between k_0 and k_1. Derive the maximum possible number of TM_z surface wave modes in a grounded dielectric structure.

3.3 Show that, for an electrically thin substrate where $h \ll \lambda_0/4\sqrt{\varepsilon_r - 1}$, only one surface wave mode exists. Also show that the surface wave number for this case, can be approximated as

$$k_{\rho 0} \approx k_0 \left[1 + \frac{1}{2}\left\{ k_0 h \left(1 - \frac{1}{\varepsilon_r} \right) \right\}^2 \right]$$

3.4 Prove that between two consecutive TM_z surface wave poles there exists one and only one TE_z surface wave pole.

3.5 The cutoff frequency for a surface wave mode is defined as $f_c = ck_{yi}/2\pi$, where c is the velocity of light in free space, k_{yi} is the surface wave number at cutoff (note that at cufoff, $k_{yi} = k_0$ and for that condition, the surface wave mode ceases to exist). Show that for a grounded dielectric structure, the TM_z mode cutoff frequencies can be solved from the equation $\tan[hk_c \sqrt{(\varepsilon_r - 1)}] = 0$. For the TE_z surface wave mode, the corresponding equation is $\cot(hk_c\sqrt{\varepsilon_r - 1}) = 0$, where $k_c = 2\pi\,(f_c/c)$.

3.6 Show that for a two-layered grounded structure, the cutoff wave numbers are given by

$$k_c \left[\frac{\sqrt{(\mu_1\varepsilon_1 - 1)}}{\varepsilon_1} \tan\alpha_1 + \frac{\sqrt{(\mu_2\varepsilon_2 - 1)}}{\varepsilon_2} \tan\alpha_2 \right] = 0$$

for TM_z mode and

$$1 - \frac{\mu_1\sqrt{(\mu_2\varepsilon_2 - 1)}}{\mu_2\sqrt{(\mu_1\varepsilon_1 - 1)}} \tan\alpha_1 \tan\alpha_2 = 0$$

for TE_z mode, where $\alpha_i = h_i k_c \sqrt{(\mu_i\varepsilon_i - 1}$, $i = 1, 2$, h_i = substrate thickness for the ith layer with (μ_i, ε_i) as the relative parameters.

3.7 A single-layered grounded dielectric structure has $h = 0.318$ cm and $\varepsilon_r = 2.56$. For a frequency of 15 GHz, how many surface wave modes would exist in the structure? Obtain the angle of incidence and the locations of apparent plane of reflection for each surface wave mode.

3.8 Derive the structure radiation factor $2F$ for a two-dielectric layered structure for the TM_z far fields when the source is located at the interface of the two dielectric layers. The dielectric thickness are h_1 and h_2 with relative parameters (ε_1, μ_1) and (ε_2, μ_2), respectively.

3.9 Derive the structure radiation function for the TE_z far fields for the structure in the Problem 3.8.

3.10 Obtain the TE_z and TM_z fields for x-directed infinitely long, electric line current source placed at the interface of two media. Both media are infinitely extended. The current distribution along x is given by $I_x = I_0 \exp(-j\beta x)$, where β is a constant. Derive the far-field patterns. Does the structure support any surface wave mode?

3.11 A planar electric surface current source is placed inside a dielectric layer, parallel to the surface of the layer. The surface current distribution can be expressed as $\mathbf{I}_s = \hat{z} \times \mathbf{H}$, where z is perpendicular to the source surface and $\mathbf{H}$ is a vector function. Prove that, if $H_z = 0$, and $\nabla \cdot \mathbf{H} = 0$, the TE_z mode will be produced by the charge accumulated at the edges of the current sheet.

3.12 A magnetic circular line current source is placed at the air-dielectric interface of a grounded dielectric structure. The magnetic line current is given by $\mathbf{M} = \hat{\phi} M_0 \cos n\phi$. Show that the z component of the electric displacement vector in the free-space region is given by

$$D_z = aM_0 \cos n\phi \int_0^\infty \frac{J_n'(k_\rho a) J_n(k_\rho \rho) \exp(-jk_z z)}{y_{\text{in}}^m(k_\rho)} k_\rho^2 \, dk_\rho$$

where $y_{\text{in}}^m(k_\rho)$ is given in (3.6) except k_y should be replaced by k_ρ. Here, a is the radius of the source and h is the substrate thickness. The interface lies on $z = 0$ plane.

3.13 Obtain the TM_z surface wave field D_z^{su} produced by the circular source in Problem 3.12. Show that, for small substrate thickness, the surface wave power can be approximated as

$$P^{\text{su}} = \frac{1}{120} \pi |M_0|^2 a^2 k_0 \frac{\{J_n'(k_{\rho 0} a)\}^2}{\left\{ 2\dfrac{h}{\varepsilon_r} + \dfrac{1}{\sqrt{k_{\rho 0}^2 - k_0^2}} \right\}}$$

3.14 Recall Problem 3.12, obtain the expression for B_z in the free-space region.

3.15 From D_z and B_z in Problems 3.12 and 3.14, obtain A_z and F_z. Use the large argument approximations for Bessel functions. Then using the saddle point method of integration, find A_z and F_z in the far-field region. Determine E_θ and E_ϕ.

Chapter 4

Characteristics of Space and Surface Wave Fields

4.1 INTRODUCTION

In this chapter we present some important characteristics of space and surface wave fields. The electromagnetic fields in a layered structure are expressed in terms of their spatial spectra. Using the complex integration method, the total field is decomposed into space and surface wave mode fields. Two surface wave modes are shown to be orthogonal to each other. Also, a surface wave field and the space wave field are found to be orthogonal. The orthogonal properties together with reciprocity are utilized to find unknown amplitudes of various surface wave modes produced by an arbitrary source. Numerical results for space and surface wave power produced by a circular magnetic current source are presented at the end.

4.2 SPACE AND SURFACE WAVE FIELDS

In this section we separate out the space and surface wave fields from a general expression of the total field in a layered structure. We consider the TM_z mode only. The procedure for the TE_z mode is quite similar.

In Chapter 3, we found that the TM_z fields produced by a planar source can generally be expressed as

$$D_z = \int_{-\infty}^{\infty}\int_{-\infty}^{\infty} \frac{f(k_x, k_y, z)}{y_{\text{in}}^m(k_x, k_y)} \exp(-jk_x x - jk_y y)\, dk_x\, dk_y \tag{4.1}$$

where the function $f(k_x, k_y, z)$ depends on the structure parameters, source type, and source intensity.

Remark. The reader should note that the function $f(k_x, k_y, z)$ is not exactly same as $f(z)$ in (3.4); it includes some more terms that are independent of z. For a one-dimensional magnetic current source,

$$f(k_x, k_y, z) = \frac{-jM_0}{2\pi} k_y \delta(k_x) f(z)$$

where $f(z)$ is defined in (3.4) for a one-layered structure.

The function $y_{\text{in}}^m (k_x, k_y)$ is independent of the source distribution and dependent solely on the structure and the source location. Using the following substitution,

$$k_x = k_\rho \cos\alpha \qquad k_y = k_\rho \sin\alpha \qquad x = \rho\cos\phi \qquad y = \rho\sin\phi$$

(4.1) is modified as

$$D_z = \int_0^{2\pi}\int_0^{\infty} \frac{f(k_\rho, \alpha, z)}{y_{\text{in}}^m (k_\rho)} \exp\{(-jk_\rho\rho)\cos(\phi - \alpha)\}\, k_\rho\, dk_\rho\, d\alpha \tag{4.2}$$

The function $f(k_\rho, \alpha, z)$ can be expanded in Fourier series, so that (4.2) becomes

$$D_z = \sum_n \int_0^{2\pi}\int_0^{\infty} \frac{f_n(k_\rho, z)}{y_{\text{in}}^m(k_\rho)} \exp(jn\alpha) \exp\{-jk_\rho\rho\cos(\phi - \alpha)\}\, k_\rho\, dk_\rho\, d\alpha \tag{4.3}$$

Now, using the Fourier-Bessel expansion of the second exponential term in the integrand of (4.3) and then performing the integration with respect to α, one obtains

$$D_z = \sum_n A_n \int_0^{\infty} \frac{f_n(k_\rho, z)}{y_{\text{in}}^m(k_\rho)} J_n(k_\rho\rho)\, k_\rho\, dk_\rho \exp(jn\phi) \tag{4.4}$$

where $A_n = (-j)^n\, 2\pi$ and $f_n\,(k_\rho, z)$ essentially is proportional to the Hankel transform of the excitation source. Equation (4.4) is a single integral representation for D_z as opposed to the double-integral form given in (4.1). The single-integral form is very convenient for separating out the space and surface wave fields. Using the following relations[1]

$$J_n(k_\rho\rho) = [H_n^{(1)}(k_\rho\rho) + H_n^{(2)}(k_\rho\rho)]/2$$

$$H_n^{(2)}(-x) = -(-1)^n H_n^{(1)}(x)$$

[1] M. Abramowitz and I.E. Stegun, *Handbook of Mathematical Functions*, National Bureau of Standards, Washington, D.C., Applied Mathematics Series 55, 1964, p. 361.

we derive the following expression for D_z:

$$D_z = \sum_n B_n \exp(jn\phi) \int_{-\infty}^{\infty} \frac{f_n(k_\rho, z)}{y_{\rm in}^m(k_\rho)} H_n^{(2)}(k_\rho\rho)\, k_\rho\, dk_\rho \tag{4.5a}$$

$$= \sum B_n \exp(jn\phi) \int_{-\infty}^{\infty} \frac{f_n(k_\rho, z)}{y_{\rm in}^m(k_\rho)} H_n^{(1)}(k_\rho\rho) k_\rho\, dk_\rho \tag{4.5b}$$

In deriving (4.5), we assummed that $f_n(-k_\rho, z) = (-1)^n f_n(k_\rho, z)$, which is always true (see Problem 4.1). The first integral form is associated with a propagating wave in the positive ρ direction. The second integral form is associated with a propagating wave in the negative ρ direction. To separate out the space and surface wave components, these integrations are performed in an alternate path as described in Chapter 3 (Figure 3.3). The integration path corresponding to the outward propagating wave is shown in Figure 4.1.

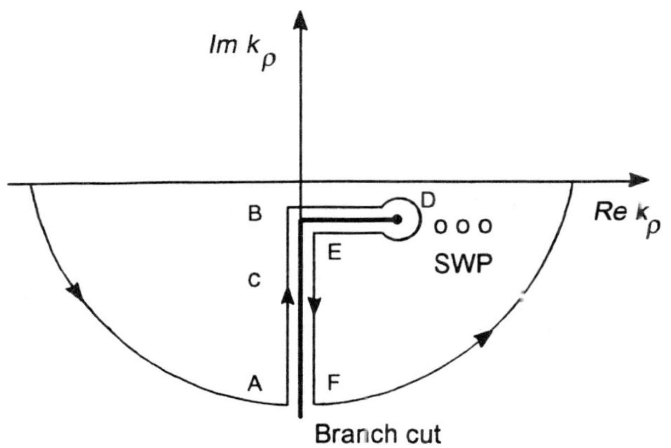

Figure 4.1 Integration path for an outward propagating wave.

Now consider the integral in (4.52). The deformed contour in Figure 4.1 consists of two circular segments and a path c around the branch cut (the branch cut is due to multivalued nature of $y_{\rm in}^m(k_\rho)$).[2] The integral vanishes on the circular segments.

[2]Note that the branch cut in Figure 4.1 is different from that in Figure 3.3. The branch cut in Figure 4.1 gives computational advantage for computing the space wave power radiated by a source. A detailed explanation is given in Section 4.5.

Ignoring the ϕ dependent term for the time being, the nth harmonic terms corresponding to the outgoing space and surface waves become

$$D_{zn}^{\text{sp}} = B_n \int_c \frac{f_n(k_\rho, z)}{y_{\text{in}}^m(k_\rho)} H_n^{(2)}(k_\rho \rho) k_\rho \, dk_\rho \tag{4.6}$$

$$D_{zn}^{\text{su}} = -2\pi j B_n \sum_i \text{res}\left\{ \frac{f_n(k_\rho, z)}{y_{\text{in}}^m(k_\rho)} H_n^{(2)}(k_\rho \rho) k_\rho, \, k_{\rho i} \right\} \tag{4.7}$$

$$= -2\pi j B_n \sum_i f_n(k_{\rho i}, z) \frac{k_{\rho i} H_n^{(2)}(k_{\rho i}\rho)}{\left.\dfrac{dy_{\text{in}}^m}{dk_\rho}\right|_{k_\rho = k_{\rho i}}} \tag{4.8}$$

where $k_{\rho i}$ is the ith surface wave number, which is equal to the ith zero of $y_{\text{in}}^m(k_\rho)$.

In a similar way, we can express the space and surface wave fields for the inward ($-\rho$ direction of propagation) traveling waves. The contour in this case, should lie on the upper half of the complex k_ρ plane, because $H_n^{(1)}(k_\rho \rho)$ vanishes on the upper semicircular segment. The TE_z space and surface wave fields are obtained similarly.

4.3 THEOREMS FOR SPACE AND SURFACE WAVE FIELDS

The space and surface wave fields exhibit some important characteristics, which will be discussed in this section.

THEOREM 4.1. *z-dependence of a surface wave field follows the source-free transmission line equation.*

Proof. As in Section 2.2.2, the two-dimensional Fourier transform of D_z for electric current excitations satisfies the following telegrapher's equations pair.

$$\frac{\partial \tilde{D}_z}{\partial z} + \varepsilon \tilde{I} = \tilde{\rho}_e$$

$$\frac{\partial \tilde{I}}{\partial z} - \frac{k_z^2}{\varepsilon} \tilde{D}_z = 0 \tag{4.9}$$

where the source term $\tilde{\rho}_e$ is the two-dimensional Fourier transform of the electric charge distribution. Referring to (4.1), $\tilde{D}_z$ is given by

$$\tilde{D}_z = \frac{f(k_x, k_y, z)}{y_{\text{in}}^m(k_x, k_y)} \tag{4.10}$$

Combining (4.9) and (4.10), one obtains

$$\frac{\partial f(k_x, k_y, z)}{\partial z} + \varepsilon \tilde{I} y_{\text{in}}^m(k_x, k_y) = \tilde{\rho}_e(k_x, k_y, z)\, y_{\text{in}}^m(k_x, k_y)$$

Now,

$$f(k_x, k_y, z) = \sum_n f_n(k_\rho, z) \exp(jn\alpha)$$

where $k_x = k_\rho \cos \alpha$, $k_y = k_\rho \sin \alpha$. Also, $y_{\text{in}}^m(k_x, k_y)$ is an explicit function of $\sqrt{(k_x^2 + k_y^2)}$, so that $y_{\text{in}}^m(k_x, k_y)$ can be expressed as a function of only one variable, k_ρ, which may be denoted as $y_{\text{in}}^m(k_\rho)$. Therefore we write

$$\sum_n \frac{\partial f_n(k_\rho, z)}{\partial z} \exp(jn\alpha) + \varepsilon \tilde{I}(k_\rho, \alpha, z)\, y_{\text{in}}^m(k_\rho) = \tilde{\rho}_e(k_\rho, \alpha, z)\, y_{\text{in}}^m(k_\rho)$$

Multiplying both sides by $\exp(-jm\alpha)$ and integrating with respect to α from 0 to 2π, one obtains

$$\frac{\partial f_m(k_\rho, z)}{\partial z} + \frac{\varepsilon}{2\pi} y_{\text{in}}^m(k_\rho) \int_0^{2\pi} \tilde{I}(k_\rho, \alpha, z) \exp(-jm\alpha)\, d\alpha$$

$$= \frac{1}{2\pi} \int_0^{2\pi} \tilde{\rho}_e(k_\rho, \alpha, z) \exp(-jm\alpha)\, d\alpha\, y_{\text{in}}^m(k_\rho)$$

or

$$\frac{\partial f_m(k_\rho, z)}{\partial z} + \varepsilon y_{\text{in}}^m(k_\rho)\, \tilde{I}_m(k_\rho, z) = \tilde{\rho}_{em}(k_\rho, z)\, y_{\text{in}}^m(k_\rho) \tag{4.11}$$

where $\tilde{I}_m$ and $\tilde{\rho}_{em}$ are the mth harmonics of $\tilde{I}$ and $\tilde{\rho}_e$, respectively. Equation (4.11) is valid for all values of k_ρ. Substituting $k_\rho = k_{\rho i}$, the righthand side of (4.11) vanishes because $y_{\text{in}}^m(k_{\rho i}) = 0$, where $k_{\rho i}$ is the ith surface wave number. Therefore,

$$\frac{\partial f_m(k_{\rho i}, z)}{\partial z} + \varepsilon \lim_{k_\rho \to k_{\rho_i}} [y_{\text{in}}^m(k_\rho) \tilde{I}_m(k_\rho, z)] = 0 \tag{4.12}$$

The reader may note that although $y_{\text{in}}^m(k_{\rho i})$ is 0 identically, the product of $y_{\text{in}}^m(k_{\rho i})$ and $\tilde{I}_m(k_{\rho i}, z)$ may not be 0 in general. This is due to the singularity of $\tilde{I}_m(k_\rho, z)$ at $k_\rho = k_{\rho i}$. Taking the mth harmonic of the second equation of (4.9), one obtains

$$\lim_{k_\rho \to k_{\rho_i}} \left\{ \frac{\partial \tilde{I}_m}{\partial z} y_{\text{in}}^m(k_\rho) \right\} - \frac{k_{zi}^2}{\varepsilon} f_m(k_{\rho i}, z) = 0 \tag{4.13}$$

Equations (4.12) and (4.13) are source-free transmission line equations, and $f_m(k_{\rho i}, z)$ represents the z-variation of the surface wave mode. This proves theorem 4.1.

This property of a surface wave mode has one very important corollary. For a given surface wave mode, the field distribution along z is unique and does not depend on the source location. Moreover, the surface wave fields are continuous along z everywhere, including in the source region. The discontinuity may occur in the transverse direction. This property will be utilized to find the surface wave field produced by a source.

THEOREM 4.2. *Surface wave modes of the same type (either TM_z or TE_z) are orthogonal along z.*

Proof. This theorem can be proven in the following way. Consider the mth harmonics of the TM_z surface wave modes. Eliminating the term $\lim_{k_\rho \to k_{\rho i}} \{y_{\text{in}}(k_\rho)\, \tilde{I}_m(k_\rho, z)\}$ from (4.12) and (4.13), one obtains the following second-order differential equation for the ith surface wave mode:

$$\frac{\partial}{\partial z} \left\{ \frac{1}{\varepsilon} \frac{\partial f_m(k_{\rho i}, z)}{\partial z} \right\} + \frac{k_{zi}^2}{\varepsilon} f_m(k_{\rho i}, z) = 0 \tag{4.14}$$

The corresponding equation associated with the jth surface wave mode is

$$\frac{\partial}{\partial z} \left\{ \frac{1}{\varepsilon} \frac{\partial f_m(k_{\rho j}, z)}{\partial z} \right\} + \frac{k_{zj}^2}{\varepsilon} f_m(k_{\rho j}, z) = 0 \tag{4.15}$$

Define a function $X(z)$ as

$$X(z) = f_m(k_{\rho j}, z) \frac{\partial f_m(k_{\rho i}, z)}{\partial z} - f_m(k_{\rho i}, z) \frac{\partial f_m(k_{\rho j}, z)}{\partial z} \tag{4.16}$$

Multiplying (4.14) by $f_m(k_{\rho j}, z)$ and (4.15) by $f_m(k_{\rho i}, z)$ and subtracting the results, one obtains

$$\frac{\partial}{\partial z} \left\{ \frac{X(z)}{\varepsilon} \right\} + \left(\frac{k_{zi}^2 - k_{zj}^2}{\varepsilon} \right) f_m(k_{\rho i}, z)\, f_m(k_{\rho j}, z) = 0 \tag{4.17}$$

Integrating (4.17) with respect to z, we obtain

$$(k_{zi}^2 - k_{zj}^2) \int \frac{f_m(k_{\rho i}, z) f_m(k_{\rho j}, z)}{\varepsilon} dz = \frac{X(z_1)}{\varepsilon(z_1)} - \frac{X(z_2)}{\varepsilon(z_2)} \tag{4.18}$$

where z_1 and z_2 are the lower and upper limits of z, respectively. Equation (4.18) is valid for a general multilayered structure with or without a ground plane. If it is an open-layered structure (like a dielectric slab with no ground plane) $z_1 = -\infty$ and $z_2 = \infty$, so that $f_m(k_{\rho i}, z)$ and its derivative both vanish at the two endpoints. This means, that the integral in (4.18) vanishes, which proves the orthogonality in the entire domain of z. If the structure has a ground plane at $z = 0$, then, the surface wave exists between $0 < z < \infty$. On the ground plane $\partial D_z/\partial z = 0$, so that $\partial f_m/\partial z = 0$ at $z = 0$. Therefore $X(z_1) = 0$. Because the surface wave decays exponentially, $X(z_2) = 0$. By virtue of (4.18), the orthogonality condition of two modes is satisfied. If the structure consists of two ground planes (a sandwich of two conductors with dielectric layers in between), then $X(z)$ will be 0 on the two conducting surfaces, so that the integral again vanishes. This kind of structure supports no space wave mode. The surface wave modes essentially are the guided modes, as in a partially filled wave guide. Therefore, for all possible layered structures, an orthogonal relationship among various TM_z surface wave modes holds; that is,

$$\int_z \frac{f_m(k_{\rho i}, z) f_m(k_{\rho j}, z)}{\varepsilon(z)} dz = 0 \tag{4.19}$$

In Figure 4.2 two consecutive surface wave mode fields supported by a dielectric coated ground plane (the TM_0 and TM_2) are plotted against z.

The preceding orthogonality is equivalent to the following relation:

$$\int_z E_{zi}^{su} \cdot H_{\phi j}^{su} \, dz = 0 \tag{4.20}$$

which shows that two different surface wave modes are uncoupled. In a similar way, the orthogonality of the TE_z surface wave modes can be shown. The orthogonal relationship for the TE_z surface wave mode is

$$\int_z \frac{g_m(k'_{\rho i}, z) \, g_m(k'_{\rho j}, z)}{\mu(z)} dz = 0 \tag{4.21}$$

where $g_m(k'_{\rho i}, z)$ denotes the z variation for the ith TE_z surface wave mode, and $k'_{\rho i}$ is the wave number associated with that mode. Recall that $k'_{\rho i}$ is the ith zero of $y_{\text{in}}^e(k_\rho)$, where $y_{\text{in}}^e(k_\rho)$ is as given in (3.20) for a grounded dielectric layered structure.

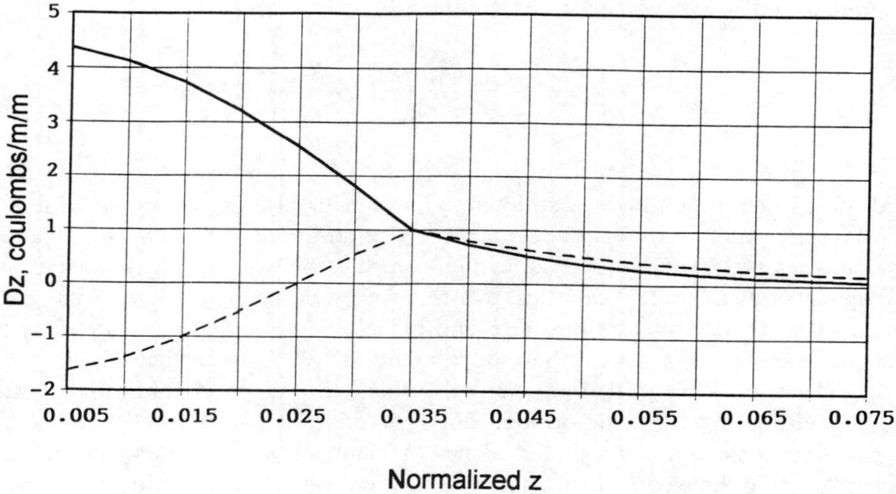

Figure 4.2 z variations of the TM_0 and TM_2 surface wave fields: ———— TM_0 mode, ---- TM_2 mode; $\varepsilon_r = 2.5$, $h = 0.7$ cm, $f = 15$ GHz.

THEOREM 4.3. *Space and surface wave fields are orthogonal along z.*

Proof. Consider the nth harmonic of TM_z space wave field. This space wave can be expressed as (see (4.6))

$$D_{zn}^{\text{sp}} = B_n \int_c \frac{f_n(k_\rho, z)}{y_{\text{in}}^m(k_\rho)} H_n^{(2)}(k_\rho \rho) k_\rho \, dk_\rho \tag{4.22}$$

where c is the contour around the branch cut (see Figure 4.1). Now, the quantity $p(z) = f_n(k_\rho, z)/y_{\text{in}}^m(k_\rho)$ satisfies the following differential equation:

$$\frac{\partial}{\partial z}\left(\frac{1}{\varepsilon}\frac{\partial p(z)}{\partial z}\right) + \frac{k_z^2}{\varepsilon} p(z) = \frac{\partial}{\partial z}\left(\frac{\tilde{\rho}_{en}}{\varepsilon}\right) \tag{4.23}$$

which can be proven upon eliminating $\tilde{I}$ from equations (4.9) and then using (4.10). In (4.23) $\tilde{\rho}_{en}$ is the nth harmonic of $\tilde{\rho}_e$. We define a function $W(z)$ as

$$W(z) = f_m(k_{\rho i}, z)\frac{\partial f_n(k_\rho, z)}{\partial z} - f_n(k_\rho, z)\frac{\partial f_m(k_{\rho i}, z)}{\partial z} \tag{4.24}$$

Then, combining (4.15), (4.23), and (4.24), we obtain

$$\frac{\partial}{\partial z}\left\{\frac{W(z)}{\varepsilon}\right\} + \frac{(k_z^2 - k_{zi}^2)}{\varepsilon} f_m(k_{\rho i}, z) f_n(k_\rho, z) = f_m(k_{\rho i}, z) \frac{\partial}{\partial z}\left(\frac{\tilde{\rho}_{en}}{\varepsilon}\right) y_{\text{in}}^m(k_\rho)$$

Integrating in the entire domain of z we obtain

$$\frac{W(z_2)}{\varepsilon(z_2)} - \frac{W(z_1)}{\varepsilon(z_1)} + (k_z^2 - k_{zi}^2) \int \frac{f_m(k_{\rho i}, z) f_n(k_\rho, z)}{\varepsilon} dz = \int f_m(k_{\rho i}, z) \frac{\partial}{\partial z}\left(\frac{\tilde{\rho}_{en}}{\varepsilon}\right) y_{\text{in}}^m(k_\rho)\, dz \tag{4.25}$$

Now, by virtue of the same argument given in Theorem 4.2, we have $W(z_1) = W(z_2) = 0$. Therefore,

$$\int \frac{f_m(k_{\rho i}, z) f_n(k_\rho, z)}{\varepsilon} dz = \int \frac{f_m(k_{\rho i}, z)}{(k_z^2 - k_{zi}^2)} \frac{\partial}{\partial z}\left(\frac{\tilde{\rho}_{en}}{\varepsilon}\right) y_{\text{in}}^m(k_\rho)\, dz \tag{4.26}$$

We now multiply both sides by $B_n[H_n^{(2)}(k_\rho \rho)k_\rho]/[y_{\text{in}}^m(k_\rho)]$ and integrate with respect to k_ρ along the contour c in Figure 4.1. The result is

$$\int_z \frac{f_m(k_{\rho i}, z)}{\varepsilon} D_{zn}^{\text{sp}}\, dz = B_n \int_z \int_c \frac{f_m(k_{\rho i}, z)}{(k_z^2 - k_{zi}^2)} \frac{\partial}{\partial z}\left(\frac{\tilde{\rho}_{en}}{\varepsilon}\right) H_n^{(2)}(k_\rho \rho) k_\rho\, dk_\rho\, dz \tag{4.27}$$

Now $(k_z^2 - k_{zi}^2) = (k_{\rho i}^2 - k_\rho^2)$. Let us first integrate the righthand side with respect to k_ρ. The integrand has no branch point and is analytic everywhere on the integration path c. The integrand vanishes on the semicircular contour. Now, because no branch point exists, the value of the integral on the path ABD of Figure 4.1 will be negative to that on the path DEF. Therefore the integral vanishes. Thus, the lefthand side of (4.27) becomes 0. This is equivalent to the following relation:

$$\int_z \frac{D_z^{\text{su}} \cdot D_z^{\text{sp}}}{\varepsilon} dz = 0 \tag{4.28}$$

which proves orthogonality between the space and surface wave fields. Similarly, the orthogonal relationship holds for the TE_z space and surface wave fields. For the TE_z mode fields, the relation is

$$\int_z \frac{B_z^{\text{su}} \cdot B_z^{\text{sp}}}{\mu} dz = 0 \tag{4.29}$$

These properties are utilized to find the surface wave fields produced by an arbitrary source. In the next section, we shall use the orthogonal relations to solve the problem of a magnetic current source.

4.4 APPLICATIONS

In this section, characteristics of the space and surface wave modes will be utilized to find the unknown surface wave fields produced by a source. The source is placed inside a layered structure with a ground plane at one end. These problems have practical importance in the area of patch antennas, where the equivalent current concept is used to obtain the outside fields.

4.4.1 Arbitrary Magnetic Current Source

Consider a magnetic current source in a layered structure as shown in Figure 4.3. An equivalent magnetic current on the radiating edge of a microstrip antenna is an example of such a source. A microstrip antenna can be of any shape, therefore, the equivalent aperture source will follow the geometry of the patch. In the following example, we shall consider an arbitrary current source and determine the surface wave mode produced by it.

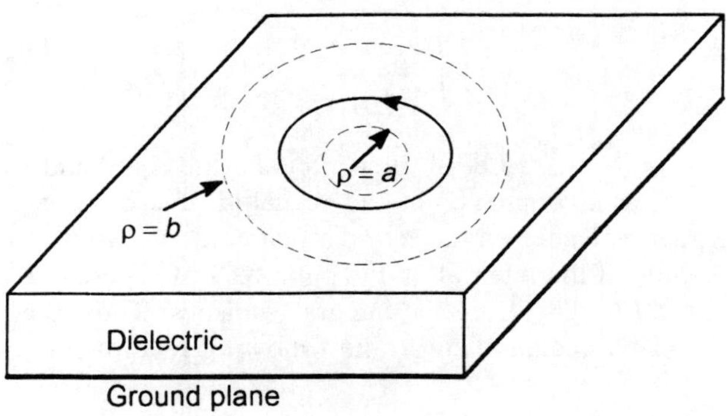

Figure 4.3 An arbitrary magnetic current source.

For an electrically small source, such as an isolated patch antenna, the surface wave fields will be cylindrical in nature as it moves away from the source. Therefore, from a convergence point of view, it is convenient to express the surface wave fields in terms of cylindrical waves.

Consider two concentric cylinders of radii a and b respectively ($b > a$), where the source lies inside the two cylinders. We shall concentrate on the TM_z surface wave fields first. The z-directed electric displacement vector can be expressed in terms of the cylindrical wave functions:

$$
\begin{aligned}
D_z &= \sum_{m,n} A_{mn} H_n^{(2)}(k_{\rho m}\rho) f_n(k_{\rho m}, z) \exp(jn\phi) + D_z^{sp} \qquad \rho > b \\
&= \sum_{mn} B_{mn} J_n(k_{\rho m}\rho) f_n(k_{\rho m}, z) \exp(jn\phi) + D_z^{sp} \qquad \rho < a
\end{aligned} \tag{4.30}
$$

where $k_{\rho m}$ is the mth zero of $y_{\text{in}}^m(k_\rho)$. The function $f_n(k_{\rho m}, z)$ is same as $f(k_\rho, z)$ given in (3.4). Note that, in the region $\rho < a$, only the Bessel function is taken because the field has to be finite at $\rho = 0$. Our next task is to determine the unknown mode amplitudes A_{mn} and B_{mn}. To do that, we invoke the reciprocity theorem. Consider an auxiliary source placed at $\rho = +\infty$, which produces the TM_{mn} surface wave mode only. The associated electromagnetic field components are obtainable from D_z^a (the superscript a stands for auxiliary), where

$$
D_z^a = H_n^{(1)}(k_{\rho m}, \rho) f_n(k_{\rho m}, z) \exp(-jn\phi)
$$

The electromagnetic field components are

$$
\begin{aligned}
H_\phi^a &= -\frac{j\omega}{k_{\rho m}} H_n^{(1)'}(k_{\rho m}, \rho) f_n(k_{\rho m}, z) \exp(-jn\phi) \\
H_\rho^a &= \frac{n\omega}{\rho k_{\rho m}^2} H_n^{(1)}(k_{\rho m}, \rho) f_n(k_{\rho m}, z) \exp(-jn\phi) \\
E_\phi^a &= \frac{-jn}{\varepsilon \rho k_{\rho m}^2} H_n^{(1)}(k_{\rho m}, \rho) \frac{\partial f_n}{\partial z} \exp(-jn\phi)
\end{aligned} \tag{4.31}
$$

For a magnetic current source, the reciprocity relation is

$$
\oiint_s (\mathbf{E}^a \times \mathbf{H}^b - \mathbf{E}^b \times \mathbf{H}^a) \cdot ds = \iiint_v (\mathbf{H}^a \cdot \mathbf{M}^b)\, dv \tag{4.32}
$$

where v is the volume inside the closed surface s.

In the preceding relation, we consider $(\mathbf{E}^b, \mathbf{H}^b)$ as the electromagnetic fields produced by the original source and $(\mathbf{E}^a, \mathbf{H}^a)$ as the fields of the auxiliary source. v is the volume of the entire annular region between two fictitious cylinders of radii a and b, respectively. Employing reciprocity, and then performing a lengthy but straightforward algebraic manipulations, we obtain

$$
\frac{4\omega}{k_{\rho m}^2} (2A_{mn} - B_{mn}) \int \frac{f_n^2(k_{\rho m}, z)}{\varepsilon(z)}\, dz = -\iiint_v \mathbf{H}^a \cdot \mathbf{M}^b\, dv \tag{4.33}
$$

In deriving this equation, the following orthogonal relationships between two surface wave modes and between space and surface wave modes are used:

$$\int_z \frac{f_n(k_{\rho m}, z)}{\varepsilon(z)} f_n(k_{\rho j}, z)\, dz = 0 \qquad m \neq j$$

$$\int_z \frac{f_n(k_{\rho m}, z)}{\varepsilon(z)} D_z^{\text{sp}}(\rho, \phi, z)\, dz = 0$$

Now let us choose another set of auxiliary fields with the following components:

$$\begin{aligned}
D_z^c &= H_n^{(2)}(k_{\rho m}, \rho) f_n(k_{\rho m}, z) \exp(-jn\phi) \\
H_\phi^c &= -\frac{j\omega}{k_{\rho m}} H_n^{(2)\prime}(k_{\rho m}, \rho) f_n(k_{\rho m}, z) \exp(-jn\phi) \\
H_\rho^c &= \frac{n\omega}{\rho k_{\rho m}^2} H_n^{(2)}(k_{\rho m}, \rho) f_n(k_{\rho m}, z) \exp(-jn\phi) \\
E_\phi^c &= \frac{-jn}{\varepsilon \rho k_{\rho m}^2} H_n^{(2)}(k_{\rho m}, \rho) \frac{\partial f_n}{\partial z} \exp(-jn\phi)
\end{aligned} \tag{4.34}$$

Using reciprocity and orthogonality once again we get

$$\frac{4\omega}{k_{\rho m}^2} B_{mn} \int \frac{f_n^2(k_{\rho m}, z)}{\varepsilon(z)}\, dz = -\iiint_v \mathbf{H}^c \cdot \mathbf{M}^b\, dv \tag{4.35}$$

The expression for B_{mn} is readily obtained from this equation. Solving A_{mn} from (4.33) and (4.35), we obtain

$$A_{mn} = -\frac{k_{\rho m}^2}{8\omega} \left[\iiint (\mathbf{H}^a + \mathbf{H}^c) \cdot \mathbf{M}^b\, dv \Big/ \int_z f_n^2(k_{\rho m}, z)/\varepsilon(z)\, dz \right] \tag{4.36}$$

This completes the determination of unknown surface wave amplitudes.

4.4.2 Circular Magnetic Current Ribbon

The preceding formulation will be used to obtain the surface wave power produced by a circular magnetic surface current ribbon.[3] The magnetic surface current is taken as

$$\mathbf{M}_b = E_0 \cos \phi \delta(\rho - a)\hat{\phi} \qquad 0 < z < h \tag{4.37}$$

This current ribbon could be the equivalent surface current on the radiating aperture of a circular patch operating in the TM_1 (the subscript 1 stands for the azimuthal variation of field underneath the patch) mode. In that case, $E_0 \cos \phi$ is the electric field on the aperture at $\rho = a$.

To determine the surface wave fields produced by the magnetic current source, we use (4.35). The auxiliary magnetic fields can be obtained using (4.31). For the present case, the magnetic field vector is given by

$$\mathbf{H}^a = \nabla \times \left\{ \hat{z} \frac{j\omega}{k_{\rho m}^2} H_1^{(1)}(k_{\rho m}\rho) f_1(k_{\rho m}, z) \cos \phi \right\} \tag{4.38}$$

which yields the ϕ component as

$$H_\phi^a = \frac{-j\omega}{k_{\rho m}} H_1^{(1)'}(k_{\rho m}\rho) f_1(k_{\rho m}, z) \cos \phi \tag{4.39}$$

The reader must note that, following the source field variation, we have taken $\cos \phi$ variation for the auxiliary field. The other auxiliary field H_ϕ^c is

$$H_\phi^c = \frac{-j\omega}{k_{\rho m}} H_1^{(2)'}(k_{\rho m}\rho) f_1(k_{\rho m}, z) \cos \phi \tag{4.40}$$

Using reciprocity, we obtain

$$\begin{aligned} A_{m1} &= \frac{jk_{\rho m}}{2} \iiint E_0 \cos^2\phi \delta(\rho - a) J_1'(k_{\rho m}\rho) f_1(k_{\rho m}, z)\rho \, d\rho \, d\phi \, dz \Big/ \int_z \frac{f_1^2(k_{\rho m}, z)}{\varepsilon(z)} dz \\ &= \frac{jk_{\rho m}a}{2} E_0 \pi J_1'(k_{\rho m}a) \frac{I_1}{I_2} \end{aligned} \tag{4.41}$$

[3]A similar procedure was used to find the fields produced by a source inside a radial waveguide. See B. Azarbar and L. Shafai, "Field Solution for Radial Waveguides with Annular Discontinuities," *IEEE Trans. on Microwave Theory and Techniques*, Vol. MTT-27, Nov. 1979, pp. 883–890.

where

$$I_1 = \int_0^h f_1(k_{\rho m}, z)\, dz = \frac{\tan(k_{1zm}h)}{k_{1zm}}$$

$$I_2 = \int_0^\infty \frac{f_m^2(k_{\rho m}, z)}{\varepsilon(z)}\, dz = \frac{1}{2\varepsilon_0}\left[\frac{1}{\varepsilon_r}\left\{h \sec^2(k_{1zm}h) + \frac{\tan(k_{1zm}h)}{k_{1zm}} + \frac{1}{\gamma_{0zm}}\right\}\right] \tag{4.42}$$

with

$$k_{1zm} = \sqrt{(k_1^2 - k_{\rho m}^2)} \qquad \gamma_{0zm} = \sqrt{(k_{\rho m}^2 - k_0^2)}$$

The TM_{m1} mode surface wave field for $\rho > a$ is (the first index m stands for the radial variation)

$$D_z^{su} = A_{m1} H_1^{(2)}(k_{\rho m}\rho) f_1(k_{\rho m}, z) \cos\phi$$

The magnetic field component in the ϕ direction is

$$H_\phi^{su} = -A_{m1} \frac{j\omega}{k_{\rho m}} H_1^{(2)'}(k_{\rho m}\rho) f_1(k_{\rho m}, z) \cos\phi$$

The r.f. (radio frequency) power carried by the TM_{m1} surface wave mode is

$$P_{m1}^{TM} = \text{real} \iiint M_\phi H_\phi^{su*} \rho\, d\rho\, d\phi\, dz$$

which simplifies to

$$P_{m1}^{TM} = \frac{\omega|E_0|^2\pi^2 a^2}{2I_2} [J_1'(k_{\rho m}a)]^2 I_1^2 \tag{4.43}$$

Note that the TM_z mode surface wave power vanishes if $J_1'(k_{\rho m}a) = 0$, which yields $k_{\rho m}a = 1.84$, 5.33, 8.54, and so forth. Because $k_{\rho m}$ is independent of the source radius (it depends entirely on the structure geometry) one can select the value of a such that the surface wave power vanishes. The smallest possible radius for the TM_{m1} surface wave mode power to vanish is $a_{\min} = 1.84/k_{\rho m}$.

Figure 4.4 shows the variations of surface wave power with the normalized source radius for the TM_{01} and TM_{21} surface wave modes. The substrate thickness is $0.35\lambda_0$ with $\varepsilon_r = 2.5$. The curve, in general, has a number of local maxima and

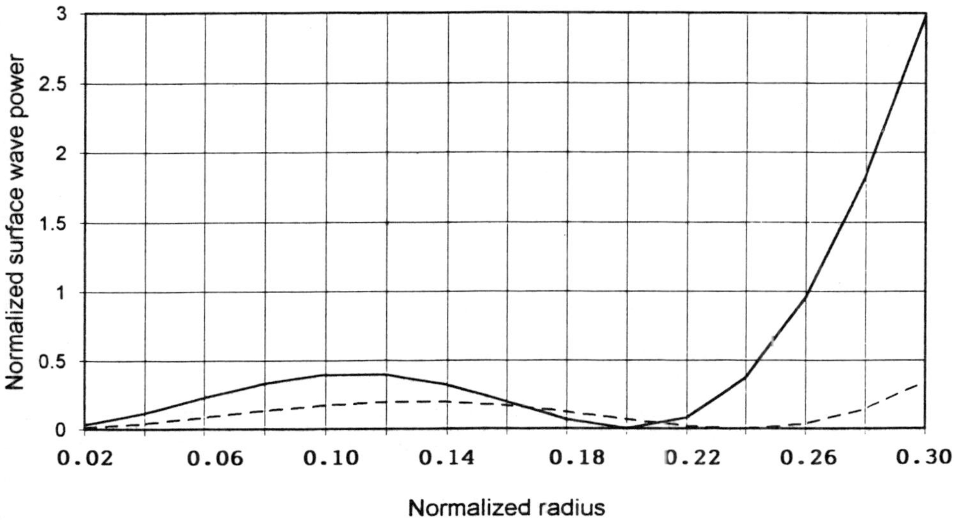

Figure 4.4 Normalized surface wave power, P_{i1}^{TM} in m-watts versus normalized source radius, a/λ_0: $E_0h = 1$ volt, $\varepsilon_r = 2.5$, $h = 0.7$ cm, $f = 15$ GHz; ________ TM_0 mode, ----- TM_2 mode.

minima. From the plot we note that a circular source with radius $a = 0.2\lambda_0$ will not excite the TM_{01} surface wave mode. Unfortunately, the *nonexistent surface wave* condition does not match with the resonant condition of a circular patch. However, annular ring patches can be designed to satisfy both conditions. Results supporting this fact has been reported in a recent paper.[4]

4.5 SPACE WAVE POWER

In Section 4.4 the reciprocity theorem was utilized to find the surface wave fields produced by an arbitrary source. We found that the use of reciprocity is very convenient to determine the surface wave fields, especially for a three-dimensional source. (This is analogous to the wave guide excitation problem. For a three-dimensional source, the use of reciprocity simplifies the problem considerably.)[5] For the space wave field, however, the use of reciprocity is not very straightforward. This is partly due to the nature of the space wave spectrum, which is continuous and complex in

[4] D.R. Jackson *et al.*, "Microstrip Patch Designs That Do Not Excite Surface Waves," *IEEE Trans. on Antennas and Propagation* (forthcoming).

[5] See, for example, R.E. Collin, *Foundations for Microwave Engineering*, New York, McGraw-Hill, 1966, pp. 183–190.

general as opposed to that of a surface wave. A relatively easier method would be by direct integration of the source, which is illustrated in this section.

Consider the magnetic current ribbon given in (4.37). This current will produce the TM$_z$ and TE$_z$ space wave fields. We shall consider the TM$_z$ field only. The TE$_z$ field can be calculated in a similar fashion.

The current ribbon considered in Section 4.4.2 is divided into a number of elementary ribbons of infinitesimal width dz_0. Each elementary ribbon can be approximated as a line source mathematically represented as

$$dM_\phi = E_0 dz_0 \cos \phi \delta(\rho - a)\delta(z - z_0) \tag{4.44}$$

where $z = z_0$ is the location of the ribbon.

The TM$_z$ mode fields are obtained from the z component of the electric displacement vector. The displacement vector component dD_z produced by the elementary magnetic current dM_ϕ is expressed in terms of the Hankel spectrum:

$$dD_z = \cos \phi \int_0^\infty d\tilde{D}_z(k_\rho, z)\, J_1(k_\rho \rho) k_\rho \, dk_\rho \tag{4.45}$$

From the spectral domain equivalent circuit, developed in Chapter 2, we can determine $d\tilde{D}_z(k_\rho, z)$ as a function of z. The spectral domain equivalent circuit is shown in Figure 4.5. Following (2.49), the current I_0 in Figure 4.5 is given by

$$\begin{aligned} I_0 &= -\int (\nabla \times \mathbf{J}_{ms} \cdot \hat{z})\, J_1(k_\rho \rho)\rho \, d\rho \\ &= -\int \frac{\partial}{\partial \rho} (\rho dM_\phi)\, J_1(k_\rho \rho) d\rho \end{aligned} \tag{4.46}$$

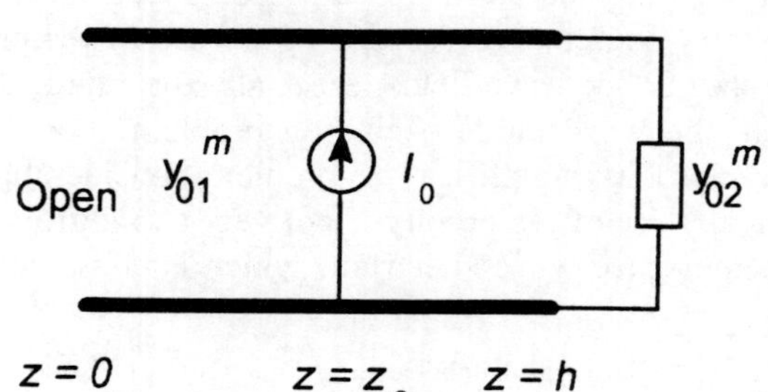

Figure 4.5 Spectral domain equivalent circuit for the TM$_z$ mode.

Performing integration by parts, we obtain

$$I_0 = E_0 \cos \phi a k_\rho J_1'(k_\rho a)\, dz_0 \tag{4.47}$$

The general expression for $d\tilde{D}_z(k_\rho, z)$ is

$$d\tilde{D}_z = \frac{I_0}{y_{in}^m(k_\rho, z_0)} f(k_\rho, z, z_0) \tag{4.48}$$

where $y_{in}^m(k_\rho, z_0)$ is the equivalent admittance at $z = z_0$, which is

$$y_{in}^m(k_\rho, z_0) = j y_{01}^m \tan(k_{1z} z_0) + y_{01}^m \frac{y_{02}^m + j y_{01}^m \tan k_{1z}(h - z_0)}{y_{01}^m + j y_{02}^m \tan k_{1z}(h - z_0)} \tag{4.49}$$

Following the procedure in Section 2.2.3, the voltage distribution function $f(k_\rho, z, z_0)$ is derived as

$$\begin{aligned} f(k_\rho, z, z_0) &= \frac{\cos(k_{1z} z)}{\cos(k_{1z} z_0)} \qquad 0 < z < z_0 \\ &= \frac{y_{01}^m \cos k_{1z}(h - z) + j y_{02}^m \sin k_{1z}(h - z)}{y_{01}^m \cos k_{1z}(h - z_0) + j y_{02}^m \sin k_{1z}(h - z_0)} \qquad z_0 < z < h \end{aligned} \tag{4.50}$$

The total displacement vector component contributed by the whole current ribbon is the superposition of individual elementary current contributions. Therefore,

$$\tilde{D}_z(k_\rho, z) = \int dD_z = E_0 a \cos \phi \int_0^h \frac{k_\rho J_1'(k_\rho a)}{y_{in}^m(k_\rho, z_0)} f(k_\rho, z, z_0)\, dz_0 \tag{4.51}$$

In the space domain, the displacement vector is

$$D_z(\rho, \phi, z) = E_0 a \cos \phi \int_0^\infty \int_0^h \frac{k_\rho J_1'(k_\rho a)}{y_{in}^m(k_\rho, z_0)} f(k_\rho, z, z_0) J_1(k_\rho \rho) k_\rho\, dz_0\, dk_\rho \tag{4.52}$$

By virtue of Section 4.2, the space wave field is given by

$$D_z^{sp}(\rho, \phi, z) = \frac{E_0 a \cos \phi}{2} \int_c \int_{z_0} \frac{k_\rho J_1'(k_\rho a)}{y_{in}^m(k_\rho, z_0)} f(k_\rho, z, z_0) H_1^{(2)}(k_\rho \rho) k_\rho\, dz_0\, dk_\rho \tag{4.53}$$

To obtain the TM$_z$ power supplied by the source, we need to know the magnetic field. The magnetic field is obtained from the usual relations:

$$\mathbf{H}^{sp} = \frac{1}{\mu}\nabla \times \mathbf{A}^{sp} \qquad \mathbf{A}^{sp} = \hat{z}A_z^{sp} \qquad \tilde{A}_z^{sp} = \frac{j\omega\mu}{k_\rho^2}\tilde{D}_z^{sp}$$

The azimuthal component of the magnetic field is

$$H_\phi^{sp} = -j\omega\frac{E_0 a\cos\phi}{2}\int_{z_0}\int_c \frac{H_1^{(2)'}(k_\rho\rho)J_1'(k_\rho a)f(k_\rho, z, z_0)}{y_{in}^m(k_\rho, z_0)}k_\rho\, dk_\rho\, dz_0 \tag{4.54}$$

The total TM$_z$ power supplied by the source is

$$P_{TM}^{sp} = -\iiint H_\phi^* M_\phi \rho\, d\rho\, d\phi\, dz$$

which becomes

$$P_{TM}^{sp} = j\omega\frac{|E_0|^2 a^2\pi}{2}\int_c\int_0^h\int_0^h \frac{J_1'(k_\rho a)}{y_{in}^m(k_\rho, z_0)}H_1^{(2)'}(k_\rho a)f(k_\rho, z, z_0)\,k_\rho\, dz\, dz_0\, dk_\rho \tag{4.55}$$

The preceding triple integral can be simplified to a single integral after performing the integrations with respect to z and z_0. The result is

$$P_{TM}^{sp} = \frac{1}{2}j\omega|E_0|^2 a^2\pi\int_c J_1'(k_\rho a)H_1^{(2)'}(k_\rho a)F(k_\rho)k_\rho\, dk_\rho \tag{4.56}$$

with

$$F(k_\rho) = \frac{h}{jk_{1z}y_{01}^m} + \frac{jy_{02}^m\tan(k_{1z}h)}{k_{1z}^2 y_{01}^m\{y_{02}^m + jy_{01}^m\tan k_{1z}h\}} \tag{4.57}$$

To facilitate numerical integration, we choose the branch cut as given in Figure 4.1. The real part of the space wave power is contributed by the segment of the path c parallel to the real k_ρ axis (BD and DE). The two segments AB and EF together do not contribute to the real part (see Problem 4.14).

Figure 4.6 shows the variation of the TM space wave power with the normalized substrate thickness. Two different substrates were considered for the analysis. For the substrate with larger dielectric constant, the space wave power varies widely with the substrate thickness. After certain value of the thickness of the sub-

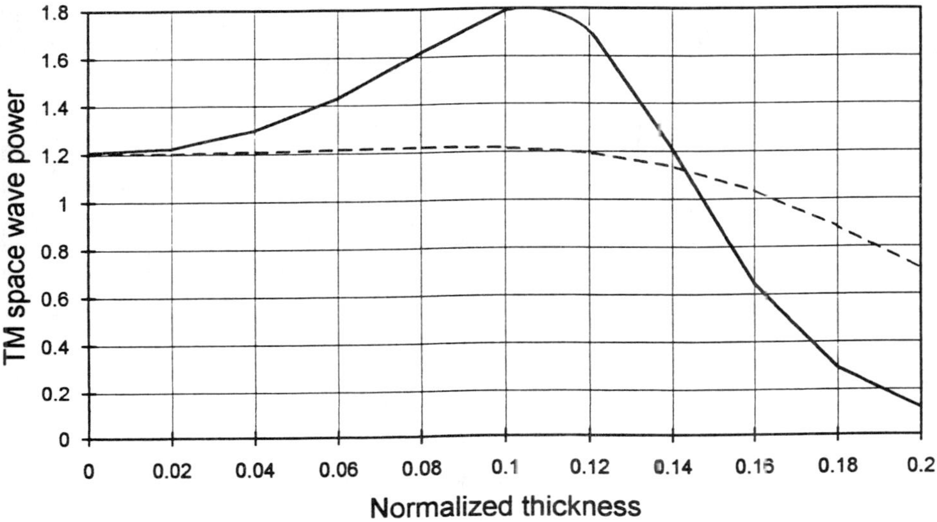

Figure 4.6 Normalized TM_z space wave power in m-watts versus normalized substrate thickness h/λ_0: $E_0 h = 1$ volt, $a = 0.3$ cm, $f = 15$ GHz; — — — —. $\varepsilon_r = 2.0$, ________ $\varepsilon_r = 4.0$.

strate, the space wave power decays rapidly. The location of the peak power depends on the dielectric constant of the substrate.

PROBLEMS

4.1 Consider a function $t(\rho, \phi) = g(\rho) \exp(jn\phi)$. The Hankel transform is defined as $\tilde{t}(k_\rho, \phi) = \exp(jn\phi) \int_0^\infty g(\rho) J_n(k_\rho \rho) \rho \, do$. Show that $\tilde{t}(-k_\rho, \phi) = (-1)^n \tilde{t}(k_\rho, \phi)$. Use this result to deduce (4.5a) and (4.5b), from (4.4).

4.2 Consider a dielectric-coated ground plane structure. The dielectric constant is ε_r, and h is the thickness of the dielectric layer. Assume that the structure supports a number of surface wave modes. By direct integration of the surface wave fields, prove the orthogonality among various TE_z or TM_z surface wave mode fields.

4.3 An x-directed magnetic surface current is placed inside a grounded dielectric structure with dielectric thickness h. The current surface is normal to the z-plane. The current is uniform along x and extended from $-\infty$ to $+\infty$. The width of the source is equal to the dielectric thickness. Using reciprocity, obtain the surface wave amplitudes in the two regions $y > 0$ and $y < 0$.

4.4 The previous problem can be solved without using reciprocity. Use continuity condition of the magnetic field and discontinuity condition of the electric field at $y = 0$, solve the unknown surface wave field amplitudes.

4.5 Figure 4.7 shows the configuration of a surface waveguide. The waveguide is excited by a ground plane transverse slot. The width of the slot is t, and the length of the slot is equal to the width of the dielectric slab. Assume that the slot field is x directed, and $E_x^{\text{slot}} = E_0$. Also, assume perfect magnetic walls at $y = 0$ and $y = b$. Determine the TM_0 surface wave fields at the two sides of the slot.

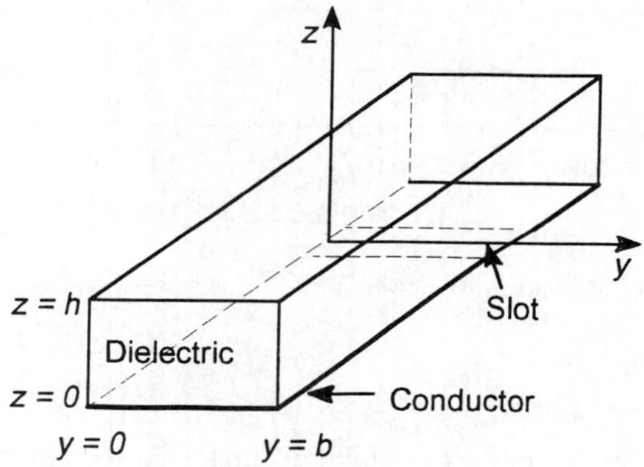

Figure 4.7 A surface waveguide made with a dielectric slab over a conducting ground plane.

4.6 Repeat Problem 4.5, if the slot length is less than b and located symmetrically. Use $E_x^{\text{slot}} = E_0 \cos[\pi/l(y - l/2)]$, where l is the slot length.

4.7 Derive expressions for D_z and B_z for an x-directed magnetic line current source in free space. The length of the current element is l. Identify the TE_z and TM_z field components. (Hints: use the free-space Green's function to obtain the electric vector potential $\mathbf{F} = F_x\hat{x}$. From $\mathbf{F}$, find E_y and E_z using $\nabla \times \mathbf{F} = -\varepsilon_0\mathbf{E}$. Use Maxwell equation to find H_z. From E_z and H_z find the TM_z and TE_z field components.)

4.8 Obtain expressions for structure radiation factors for the TE_z and TM_z fields for a grounded dielectric structure. The dielectric thickness is h with dielectric constant ε_r. Use the solutions to Problem 4.7 to find the radiation pattern of the source when the source is located on the grounded dielectric structure. Using far-field integration, determine the space wave power emanating from the source.

4.9 Figure 4.8 shows a partially dielectric-filled waveguide whose x and z dimensions are a and b, respectively. The dielectric thickness is h. The waveguide is excited by a magnetic line current source given by

$$\mathbf{M} = \hat{x}M_0 \sin\frac{\pi x}{a}\, \delta(y)\, \delta(z - h)$$

Express D_z as

$$D_z(x, y, z) = \sin\frac{\pi x}{a} \int_{-\infty}^{\infty} \tilde{D}_z(k_y, z) \exp(-jk_y y)\, dk_y$$

(a) Obtain the spectral domain equivalent circuit for the TM_z mode.
(b) Determine $y_{\text{in}}^m(k_y)$ and find the expression for $\tilde{D}_z(k_y, z)$.
(c) Show that the integral for D_z has a number of singularities but no branch point. (Note that each singular point corresponds to a guided mode).

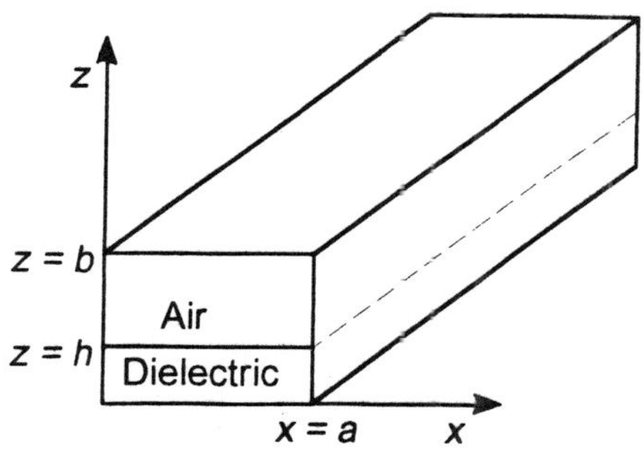

Figure 4.8 A partially dielectric-filled rectangular waveguide.

4.10 In Problem 4.9, use the contour integration technique (residue calculus) to express $D_z(x, y, z)$ in a series form (in the region $y > 0$). Show analytically that, in the series, the terms are mutually orthogonal in z.

4.11 Assume k_{y0} is the lowest order zero of $y_{\text{in}}^m(k_y)$ in Problem 4.9. Express the TM_z field components associated with that pole. Assuming k_{y0} real, determine the real power coupled to that mode.

4.12 Obtain the TE_z guided modes supported by the structure in Problem 4.9.

4.13 The partially filled waveguide in Figure 4.8 is excited by a thin probe located at $x = a/2$. The probe is parallel to the z-axis. Assume uniform current distribution on the probe. Determine the power coupled to the lowest order TM_z mode.

4.14 Deduce (4.56) from (4.55). Show that, with reference to Figure 4.1, the total contribution to real part of P_{TM}^{sp} contributed by the segments AB and EF is 0. (Hints: Note that due to the presence of the branch cut, the values of y_{02}^{m} on AB and EF are 180° out of phase. Use this result to show that the real part contributed by the segment AB is negative to that contributed by the segment EF.)

Chapter 5
Analysis of Microstrip Antennas and Arrays

5.1 INTRODUCTION

In this chapter we present the spectral domain analysis of printed radiating structures. They include microstrip antenna elements, printed array antenna, and printed screen polarizer. For microstrip antenna analysis, we use two different models: the edge admittance model, and the patch current model. The edge admittance model is applicable for regular geometrically shaped patches, such as rectangular, circular, or elliptical. We demonstrate the model with a specific example of a probe-fed circular patch antenna. The edge admittance of the patch, which is an essential part of the model, is obtained using the spectral domain analysis. Next, the patch current model is demonstrated through an example of a slot-fed printed dipole. The patch current model is commonly used in the moment method formulation of a microstrip antenna.

An analysis of infinite array sources on dielectric-coated ground plane is also presented in this chapter. An equivalent circuit of magnetic line current sources is shown that gives physical insight to the scan blindness problem in a practical microstrip phased-array antenna. Quantitative analyses of grating lobes, scan blindness, and related mismatch problems are also presented.

The chapter ends with an analysis of printed screen polarizer. For normal incident fields, a polarizer is electrically equivalent to parallel plate waveguides. The orientation of a waveguide depends on the direction of the polarization vector. The equivalent circuit of a typical polarizer is shown. In the equivalent circuit, a strip screen is equivalent to a lumped susceptance. We present the formulation for this lumped susceptance, including the effects of the adjacent dielectric media. An analysis of multilayered screen polarizer is also outlined. Finally, electrical performance of a two-layered polarizer is shown.

5.2 THE EDGE ADMITTANCE MODEL

The edge admittance model is a very simple but fairly accurate model for patch antenna analysis. This model is applicable to separable patch geometries.[1] Most of the useful patch shapes belong to this category. In this model, the wave equation is solved in the region between the patch metallization and the ground plane. The impedance wall boundary condition is imposed to incorporate the effect of r.f. radiation. From the field solution underneath the patch, the input impedance and other related characteristics are determined. In the following subsections we will explain the model with a specific example of a probe-fed circular patch antenna.

5.2.1 Fields Underneath a Circular Patch

In this subsection, we solve the electromagnetic fields underneath the patch metallization. Figure 5.1 shows a probe-fed circular patch antenna. To analyze a circular structure, the cylindrical coordinate system is most suitable. In the cylindrical coordinate system, the wave equation is separable. In other words, a solution of the wave equation can be expressed as a product of three *single-variable* functions. The independent variables in this case are ρ, ϕ, and z.

To analyze a patch antenna we usually assume that the field quantities underneath the patch are independent of z. This is primarily due to the small substrate thickness ($h < 0.05\lambda_0$, λ_0 being the wavelength in free space). The magnetic field will have no z component, because this component cannot be independent of z inside two electric conducting planes. Therefore, only the TM_z mode will exist underneath the patch. To obtain the complete field components, one should find the z component of the electric field, E_z, first. Under the source-free condition, this electric field component satisfies the following wave equation:

$$\nabla_t^2 E_z + k_1^2 E_z = 0 \tag{5.1}$$

where

$$k_1^2 = \omega^2 \mu \varepsilon_0 \varepsilon_r \qquad \nabla_t^2 = \frac{\partial^2}{\partial x^2} + \frac{\partial^2}{\partial y^2} \tag{5.2}$$

In cylindrical coordinate system, (5.2) becomes

---

[1]Separable patch geometries are those for which the wave equation is separable in a coordinate system suitable for the patch shape. For more details, see A. K. Bhattacharyya and R. Garg, "Generalized Transmission Line Model for Microstrip Patches," *IEE Proceedings*, Part H, Vol. 132, 1985, pp. 93–98.

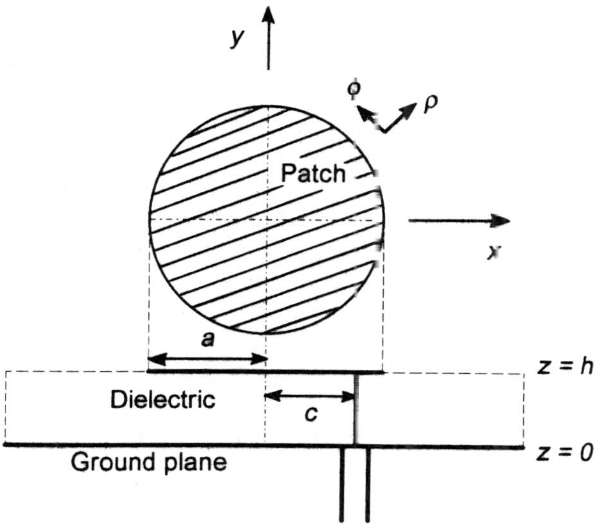

Figure 5.1 A probe-fed circular patch antenna.

$$\frac{1}{\rho}\frac{\partial}{\partial\rho}\left(\rho\frac{\partial E_z}{\partial\rho}\right)+\frac{1}{\rho^2}\frac{\partial^2 E_z}{\partial\phi^2}-k_1^2E_z=0 \tag{5.3}$$

Using the *separation of variables* method, the general solution for E_z can be found as

$$E_z=\sum_n\begin{Bmatrix}J_n(k_1\rho)\\Y_n(k_1\rho)\end{Bmatrix}\cdot\begin{Bmatrix}\cos n\phi\\\sin n\phi\end{Bmatrix} \tag{5.4}$$

where the symbol

$$\begin{Bmatrix}f(x)\\g(x)\end{Bmatrix}$$

denotes a linear combination of $f(x)$ and $g(x)$. In (5.4) J_n and Y_n represent the Bessel and Neumann functions, respectively, of order n. In (5.4) n could be integer or real number, in general. For the present geometrical shape (circular), the field has to be periodic in ϕ with a period 2π, so that n must take integral values. We write the field solutions as

$$E_z = \sum_{n=0}^{\infty} A_n J_n(k_1\rho) \cos n\phi \qquad \rho < c \tag{5.5}$$

$$E_z = \sum_{n=0}^{\infty} [B_n J_n(k_1\rho) + C_n Y_n(k_1\rho)] \cos n\phi \qquad c < \rho < a \tag{5.6}$$

Note that only a cosine variation is taken for E_z. Because the feed is located at $\phi = 0$, E_z should be symmetrical with respect to $\phi = 0$ plane. Furthermore, to make the field finite at $\rho = 0$, the Neumann function is excluded in the region $\rho < c$. The electric field component E_z is continuous at $\rho = c$. This gives

$$A_n J_n(k_1 c) = [B_n J_n(k_1 c) + C_n Y_n(k_1 c)] \tag{5.7}$$

We now apply the following discontinuity condition at the source location:

$$H_\phi(c+) - H_\phi(c-) = I_s \tag{5.8}$$

where I_s is the z-directed surface current source and H_ϕ represents the azimuthal component of the magnetic field vector. For a thin probe, I_s can be expressed as

$$I_s = \frac{I_p}{c} \delta(\phi) \tag{5.9}$$

where I_p is the total current on the probe. To obtain H_ϕ, we use Maxwell's equation $\nabla \times \mathbf{E} = -j\omega\mu\mathbf{H}$. This yields

$$H_\phi = \frac{1}{j\omega\mu} \frac{\partial E_z}{\partial \rho} \tag{5.10}$$

Substituting E_z from (5.5) and (5.6) in (5.10), we obtain

$$\frac{-jk_1}{\omega\mu} \sum_{n=0}^{\alpha} [B_n J_n'(k_1 c) + C_n Y_n'(k_1 c) - A_n J_n'(k_1 c)] \cos n\phi = I_s = \frac{I_p}{c} \delta(\phi) \tag{5.11}$$

We now expand the righthand side of (5.11) in a Fourier cosine series. This yields

$$\frac{-jk_1}{\omega\mu} [B_n J_n'(k_1 c) + C_n Y_n'(k_1 c) - A_n J_n'(k_1 c)] \pi\sigma_n = \frac{I_p}{c} \tag{5.12}$$

where $\sigma_0 = 2$, and $\sigma_n = 1$ for $n \neq 0$.

Another equation can be formed from the admittance wall boundary condition at $\rho = a$. The equivalent admittance is the representative of the radiated power and stored energy in the fringing fields. Therefore, at $\rho = a$, we can write

$$\frac{-aH_\phi}{E_z} = y_n \tag{5.13}$$

where y_n is the wall admittance for the TM_n mode. The negative sign on the lefthand side of (5.13) is taken to make the real part of y_n positive. Substituting the expressions for H_ϕ and E_z in (5.13), we obtain

$$\frac{jk_1 a}{\omega\mu} \frac{B_n J_n'(k_1 a) + C_n Y_n'(k_1 a)}{B_n J_n(k_1 a) + C_n Y_n(k_1 a)} = y_n \tag{5.14}$$

Equations (5.7), (5.12), and (5.14) can be solved to obtain A_n, B_n, and C_n. The input impedance is corresponding to the TM_n mode is

$$Z_{\text{in}}^{(n)} = \frac{-hE_z^{(n)}(\rho = c)}{I_p} = -h\frac{A_n J_n(k_1 c)}{I_p}$$

The total impedance is

$$Z_{\text{in}} = \sum_n Z_{\text{in}}^{(n)} \tag{5.15}$$

To solve for A_n, the wall (edge) admittance, y_n, is to be known. The wall admittance of a circular patch will be considered next.

5.2.2 Edge Admittance of a Circular Patch

There are two different methods to determine the edge admittance of a patch antenna. In the first method, the effects of the dielectric substrate on the radiated fields are ignored and the free-space Green's function is used to find the outside fields.[2] The second method is more general and includes the effects of dielectric substrate. The first method is very simple to use and fairly accurate for electrically thin ($h < 0.02\lambda_0$) and low permittivity substrates ($\varepsilon_r < 2.5$). However, for a thicker substrate, the first

---

[2]A. K. Bhattacharyya and R. Garg, "Self and Mutual Admittance Between Two Concentric, Coplanar, Circular Radiating Current Sources," *IEE Proceedings*, Part H, Vol. 131, 1984, pp. 217–219.

method becomes inaccurate. Here we will present the second method, which includes the effect of dielectric substrate on the edge admittance.

To determine the outside field emanating from a patch antenna, the principle of equivalence is employed. The patch aperture is replaced by its equivalent magnetic current. Let $\hat{z}E_a \cos n\phi$ be the electric field for the TM_n mode on the aperture at $\rho = a$. The equivalent magnetic surface current is

$$\mathbf{M} = E_a \cos n\phi(\hat{z} \times \hat{\rho}) = \hat{\phi}E_a \cos n\phi \qquad 0 < z < h \tag{5.16}$$

The magnetic volume current density will be

$$\mathbf{J}_m = E_a \cos n\phi\delta(\rho - a)\hat{\phi} \qquad 0 < z < h \tag{5.17}$$

This magnetic current is immersed in the dielectric substrate. To simplify the analysis, we approximate the surface current as a line current and place the line current at the air-dielectric interface. The line current can be expressed as

$$\mathbf{J} = E_a h \cos n\phi\delta(\rho - a)\, \delta(z - h)\hat{\phi} \tag{5.18}$$

This current source will produce the TE_z and TM_z fields. For the TM_z field we solve for D_z and for the TE_z fields, we solve for B_z. In a multilayered dielectric media, D_z satisfies the following equations (see Section 1.4):

$$\nabla^2 D_z + k^2 D_z = \frac{-\partial \varepsilon}{\partial z} (\nabla_t \cdot \mathbf{E}_t) + \varepsilon(\nabla \times \mathbf{J}) \cdot \hat{z} \tag{5.19}$$

$$\frac{\partial D_z}{\partial z} + \varepsilon(\nabla_t \cdot \mathbf{E}_t) = 0 \tag{5.20}$$

To solve for D_z, we express D_z in terms of its Hankel transform:

$$D_z = \cos n\phi \int_0^\infty \tilde{D}_z(k_\rho, z)\, J_n(k_\rho \rho) k_\rho\, dk_\rho \tag{5.21}$$

The cosine azimuthal variation for D_z is taken to comply with the source variation along ϕ. We also express $\nabla_t \cdot \mathbf{E}_t$ in terms of Hankel transforms:

$$\nabla_t \cdot \mathbf{E}_t = \cos n\phi \int_0^\infty I(k_\rho, z)\, J_n(k_\rho \rho) k_\rho\, dk_\rho \tag{5.22}$$

We now substitute (5.21) and (5.22) in (5.19) and (5.20), and with very little algebraic manipulation the following telegrapher's equations are deduced:

$$\frac{\partial \tilde{D}_z}{\partial z} + \varepsilon I = 0 \tag{5.23}$$

$$\frac{\partial I}{\partial z} = \frac{k_z^2}{\varepsilon} \tilde{D}_z + I_0 \delta(z - h) \tag{5.24}$$

where

$$I_0 = \frac{1}{\cos n\phi} \int \nabla \times \mathbf{J} \cdot \hat{z} \, J_n(k_\rho \rho) \rho \, d\rho$$

Substituting expressions for **J** and performing the integration by parts, we obtain

$$I_0 = E_a h k_\rho a J'_n(k_\rho a) \tag{5.25}$$

Equations (5.23) and (5.24) are a telegrapher's equations pair. Therefore, in the Hankel domain, the equivalent circuit would consist of transmission line sections. See Figure 5.2. The transmission line parameters are $y_{01}^m = jk_{1z}/\varepsilon_0\varepsilon_r$, $y_{02}^m = jk_{0z}/\varepsilon_0$.

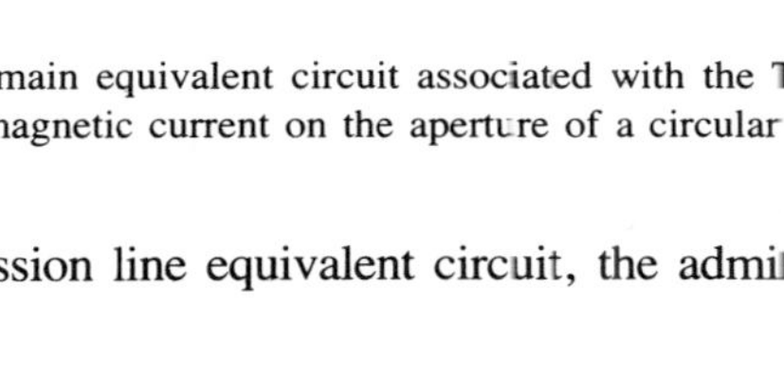

Figure 5.2 The spectral domain equivalent circuit associated with the TM_z mode fields generated by the equivalent magnetic current on the aperture of a circular patch.

Using the transmission line equivalent circuit, the admittance at the source point ($z = h$) is obtained as

$$y_{\text{in}}^m = jy_{01}^m \tan(k_{1z}h) + y_{02}^m \tag{5.26}$$

The line voltage at $z = h$ should be $V_0 = \tilde{D}_z(k_\rho, h) = I_0/y_{\text{in}}^m$. The voltage distribution in the region $0 < z < h$ will be $\tilde{D}_z(k_\rho, z) = V_0[\cos(k_{1z}z)/\cos(k_{1z}h)]$. Therefore, in the region $0 < z < h$, D_z is given by

$$D_z(\rho, \phi, z) = E_a ah \cos n\phi \int_0^\infty \frac{J_n'(k_\rho a) J_n(k_\rho \rho) \cos(k_{1z} z)}{y_{\text{in}}^m \cos(k_{1z} h)} k_\rho^2 \, dk_\rho \tag{5.27}$$

For the TE$_z$ mode fields, we need the magnetic displacement vector component B_z. Using a similar procedure we can calculate this quantity. The equivalent circuit for B_z can be obtained using Table 2.1. The expression for B_z in the region $0 < z < h$ becomes

$$B_z = -\sin n\phi \frac{\mu_1 n E_a h}{j\omega} \int_0^\infty \frac{J_n(k_\rho a) J_n(k_\rho \rho) \sin(k_{1z} z)}{z_{\text{in}}^e k_{1z} \cos(k_{1z} h)} k_\rho \, dk_\rho \tag{5.28}$$

where

$$z_{\text{in}}^e = \frac{1}{y_{02}^e} + \frac{j}{y_{01}^e} \tan(k_{1z} h)$$

$$y_{01}^e = \frac{jk_{1z}}{\mu_1} \qquad y_{02}^e = \frac{jk_{0z}}{\mu_0}$$

From D_z and B_z we find A_z and F_z, respectively. The magnetic field component H_ϕ in terms of A_z and F_z is

$$H_\phi = -\frac{1}{\mu} \frac{\partial A_z}{\partial \rho} + \frac{1}{j\omega\mu\varepsilon\rho} \frac{\partial^2 F_z}{\partial \phi \partial z}$$

The expression for H_ϕ at $z = h$ becomes

$$H_\phi(\rho) = -j\omega \cos n\phi E_a ah \int_0^\infty \frac{J_n'(k_\rho a) J_n'(k_\rho \rho)}{y_{\text{in}}^m} k_\rho \, dk_\rho - \frac{\cos n\phi E_a h n^2}{j\omega\rho} \int_0^\infty \frac{J_n(k_\rho a) J_n(k_\rho \rho)}{z_{\text{in}}^e k_\rho} dk_\rho \tag{5.29}$$

The wall admittance, by definition is

$$y_n = -\frac{aH_\phi(a)}{E_z(a)} = j\omega a^2 h \int_0^\infty \frac{[J_n'(k_\rho a)]^2}{y_{\text{in}}^m} k_\rho \, dk_\rho - \frac{jhn^2}{\omega} \int_0^\infty \frac{J_n^2(k_\rho a)}{z_{\text{in}}^e k_\rho} dk_\rho \tag{5.30}$$

These integrals are computed numerically. The real parts of the integrals converge rapidly, although the imaginary parts do not. This is because the surface current is approximated by a line current that has infinite susceptance value. To circumvent the nonconvergence problem, we consider that the magnetic current source is a surface current lying on the surface $z = h$. The width of the surface current is taken to be the same as the substrate thickness. With this assumption, the modified expression for y_n becomes

$$y_n = \frac{j\omega}{h}\int_0^{\infty}\frac{\left[\int_a^{a+h} J_n'(k_\rho \rho)\rho\, d\rho\right]^2}{y_{\text{in}}^m} k_\rho\, dk_\rho - \frac{jn^2}{\omega h}\int_0^{\infty}\frac{\left[\int_a^{a+h} J_n(k_\rho \rho)\, d\rho\right]^2}{z_{\text{in}}^e k_\rho}\, dk_\rho \tag{5.31}$$

Numerical evaluations of the integrals in (5.31) are straightforward in the region $0 < k_\rho < k_0$. However, in the region $k_0 < k_\rho < k_1$, the first integral has at least one pole associated with the lowest order TM_z surface wave mode. The second integral, however, will have no pole for thin substrates. In the neighborhood of the singular point, the integral is evaluated using the singularity extension technique described in Appendix A2. The residue of the integral at the singular point yields the contribution of the surface wave mode to the total wall conductance.

In Figure 5.3 the wall conductance of a circular patch is plotted. The wall conductance has two components: the space wave conductance and the surface wave conductance. The surface wave conductance is responsible for the surface wave power,

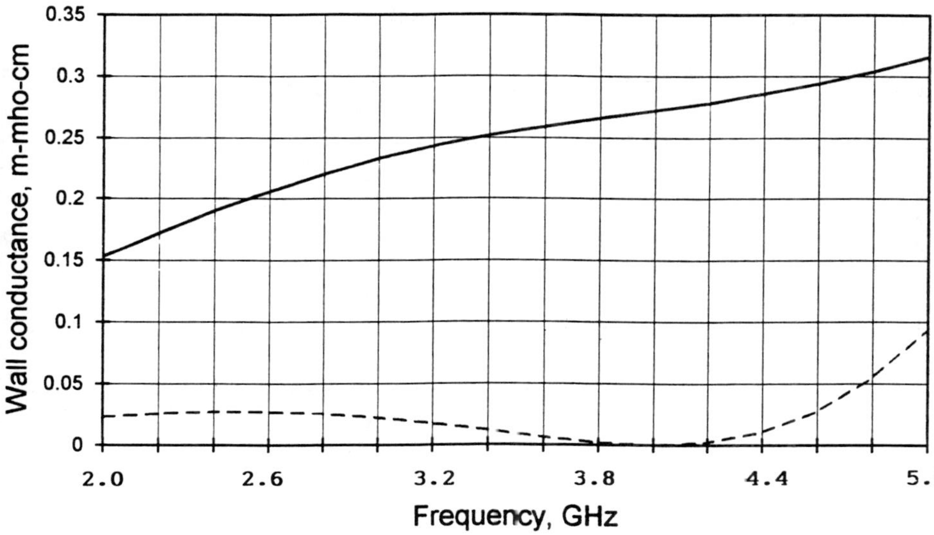

Figure 5.3 Edge conductance of a circular patch versus frequency: $a = 2$ cm, $h = 0.318$ cm, $\varepsilon_r = 2.65$; —— space wave, ---- surface wave.

and the space wave conductance is responsible for the radiated power. The radiation efficiency is defined as the fraction of the total power that goes to the radiated field. In terms of the wall conductance, the radiation efficiency would be

$$\text{Radiation efficiency} = \frac{g_{sp}}{g_{sp} + g_{su}} \tag{5.32}$$

where g_{sp} and g_{su} are the space and surface wave conductances, respectively.

Note that the present wall admittance formulation does not include the effect of the patch conductor. In fact, the patch conductor has a considerable effect on the wall susceptance part, due to the electric charge that accumulates on the edge of the conductor. The accumulated charge produces reactive energy, effectively increasing the susceptance value. The contribution of the electric charge is equivalent to a capacitance. For a small substrate thickness, this capacitance may be approximated as the fringing field capacitance due to static charge. The fringing field capacitance was first determined by Kirchhoff using conformal mapping.[3] The expression for the capacitance is $c_f = \varepsilon_r \sqrt{\mu \varepsilon_0}\, a \Delta a / \eta$, with $\Delta a = h/[\ln(\pi a/2h) + 1.7726]/\pi\varepsilon_r$.

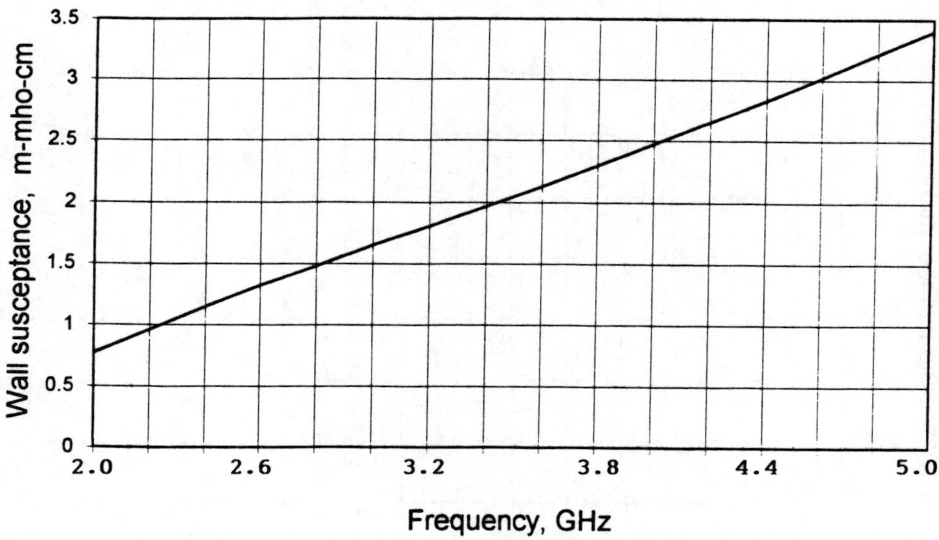

Figure 5.4 Edge susceptance of a circular patch antenna: $a = 2$ cm, $h = 0.318$ cm, $\varepsilon_r = 2.65$.

---

[3]G. Kirchhoff, *Monatsb Deutsch Akad. Wiss.* (Berlin), 1877, p. 144.

The total wall admittance, including the effect of the patch conductor will be

$$y_w^n = j\omega c_f + y_n \tag{5.33}$$

where y_n is given in (5.31). In Figure 5.4 the wall susceptance of a circular patch antenna is plotted against frequency. Numerically, we find that the wall susceptance value is dominated by the contribution of the accumulated charge at the conductor edge. The equivalent magnetic current contributes very little, especially for a small substrate thickness.

5.2.3 Input Impedance

Equations (5.15) and (5.33) were used to compute the input impedance of a circular patch. The impedance is computed in the vicinity of the TM_1 mode resonant frequency, because that is the most useful mode for antenna applications. About five modes were taken to compute the input impedance. Figure 5.5 shows the input impedance variation of a typical circular patch antenna. The input impedance is a function of the feed location. The resonant resistance increases as the feed moves away from the center of the patch. The location of the feed is usually adjusted to obtain a perfect match with the coaxial feed.

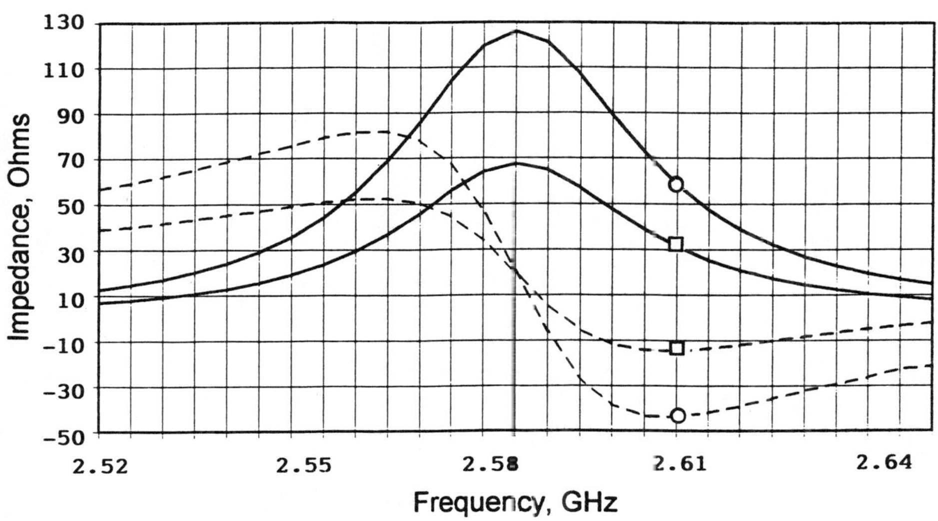

Figure 5.5 Input impedance of a probe-fed circular patch antenna versus frequency: a = 2 cm, h = 0.159 cm, ε_r = 2.65; —O— R (c = 0.7 cm), —□— R (c = 0.5 cm), --O-- X (c = 0.7 cm), --□-- X (c = 0.5 cm).

5.3 THE PATCH CURRENT MODEL

In the preceding section, we demonstrated the edge admittance model of a patch antenna. In another useful model for patch analysis, the patch conductor is replaced by the induced electric current on it. This model, commonly used for the moment method formulation of a patch, is versatile in that it can be used for any patch shape. The edge admittance model, however, is limited to regular shaped patch, where the eigen functions (to expand the fields underneath the patch) are available in closed forms.

In this section, we will demonstrate the patch current model through a simple example of an aperture-coupled dipole antenna. Figure 5.6 shows the geometry of a microstrip dipole antenna. The dipole is excited by a slot on the ground plane. The slot field will induce electric current on the dipole conductor. The induced current is responsible for radiation. The model essentially involves the following three major steps.

1. To express the fields of the dipole for assumed current distribution on the dipole surface;
2. To determine the induced current on the dipole conductor;
3. To determine the input impedance.

We will consider them one by one.

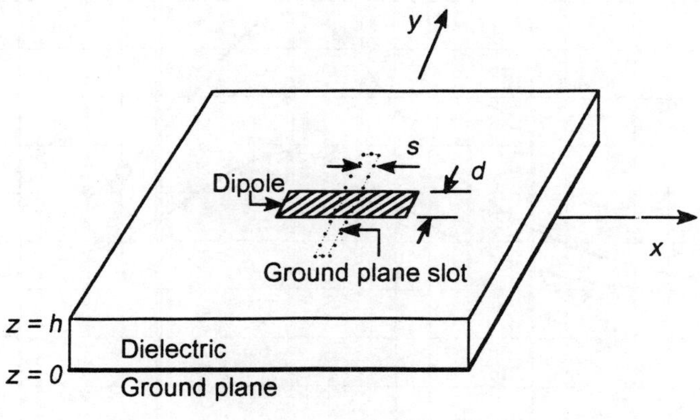

Figure 5.6 The structure of a slot-fed dipole antenna. a = dipole length, l = slot length.

5.3.1 Fields Produced by a Dipole Source

In this section we will find the electromagnetic fields produced by the unknown induced current on the dipole surface. The induced current distribution on the dipole is assumed as

$$I_x = \frac{I_d}{d} \cos\frac{\pi x}{a}$$

where I_d is an unknown, to be found. The reader may note that this is a first-order approximation for the dipole current distribution. For rigorous calculation, we should express the dipole current in terms of a series of cosines functions with unknown amplitudes. In the present analysis, however, we will restrict ourselves to only one term approximation. Variables a and d are the length and width of the dipole, respectively. Using Table 2.1, we can construct the spectral domain equivalent circuits for the TM_z and TE_z modes. From the equivalent circuit of the TM_z mode, we obtain the Fourier spectrum of D_z as

$$\tilde{D}_z = -\tilde{\rho}_{se} \frac{y_{02}^{m} \cos(k_{1z}z)}{y_{in}^{mh} \cos(k_{1z}h)} \qquad 0 < z < h \tag{5.34}$$

where y_{in}^{mh} is the input admittance at $z = h$, which is $y_{in}^{mh} = y_{02}^{m} + jy_{01}^{m} \tan(k_{1z}h)$.

In (5.34), $\tilde{\rho}_{se}$ is the spectrum of the electric surface charge density ρ_{se}. Through continuity equation, $\tilde{\rho}_{se}$ can be expressed as

$$\tilde{\rho}_{se} = \frac{k_x}{\omega} \tilde{I}_x$$

where $\tilde{I}_x$ is the Fourier transform of I_x. $\tilde{I}_x$ has a closed form expressions as

$$\tilde{\mathbf{I}}_x = \frac{-I_d}{2\pi a} \left\{ \frac{\sin(k_y d/2)}{k_y d/2} \right\} \left\{ \frac{\cos(k_x a/2)}{k_x^2 - (\pi/a)^2} \right\}$$

The Fourier spectrum for B_z is obtained from the TE_z mode spectral domain equivalent circuit. The final expression for $\tilde{B}_z$ is

$$\tilde{B}_z = \frac{jk_y \tilde{I}_x \sin(k_{1z}z)}{y_{in}^{eh} \sin(k_{1z}h)} \qquad 0 < z < h \tag{5.35}$$

where

$$y_{in}^{eh} = \frac{jk_{0z}}{\mu_0} + \frac{k_{1z}}{\mu_0 \mu_r} \cot(k_{1z}h)$$

is the equivalent admittance at $z = h$ location. From $\tilde{D}_z$ and $\tilde{B}_z$, we find A_z and F_z, respectively. As usual the electromagnetic field components are obtained from A_z and F_z. With reference to the present problem, the x component of the electric field and the y component of the magnetic field are of interest. In the spectral domain, they are

$$\tilde{E}_x^d = \frac{-\omega k_y}{k_\rho^2}\tilde{B}_z - \frac{jk_x}{\varepsilon_0\varepsilon_r k_\rho^2}\frac{\partial \tilde{D}_z}{\partial z} \tag{5.36}$$

$$\tilde{H}_y^d = \frac{-\omega k_x}{k_\rho^2}\tilde{D}_z - \frac{jk_y}{\mu_0\mu_r k_\rho^2}\frac{\partial \tilde{B}_z}{\partial z} \tag{5.37}$$

where $k_\rho^2 = k_x^2 + k_y^2$. The superscript d stands for the fields contributed by the dipole current.

5.3.2 Amplitude of the Dipole Current

To determine the unknown dipole current amplitude I_d, the electric field boundary condition on the dipole surface is enforced. The total tangential electric field must vanish on the dipole surface. This yields

$$E_x^d + E_x^s = 0 \tag{5.38}$$

where E_x^s stands for the electric field produced by the primary source, in this case, the ground plane slot. We now multiply equation (5.38) by I_x and integrate over the dipole surface. This yields

$$\iint E_x^d I_x \, dx \, dy + \iint E_x^s I_x \, dx \, dy = 0 \tag{5.39}$$

The first integral is the self-reaction of the dipole current and the second integral is the mutual reaction between the dipole and the slot. The first integral is proportional to I_d^2 and the second integral is proportional to $I_d V_s$, where V_s is the slot voltage. Mathematically, we can write (5.39) as $(I_d)^2$ Int1 + $(I_d V_s)$ Int2 = 0, yielding

$$I_d = -V_s \, \text{Int2}/\text{Int1} \tag{5.40}$$

The term Int1 is given by Int1 = $(1/I_d^2) \iint E_x^d I_x \, dx \, dy$. Upon application of Parseval's theorem, this integral can be expressed as,

$$\text{Int1} = \frac{4\pi^2}{I_d^2}\int_{-\infty}^{\infty}\int_{-\infty}^{\infty} \tilde{E}_x^d\big|_{z=h}\tilde{I}_x \, dk_x \, dk_y \tag{5.41}$$

To evaluate the second integral, the reciprocity theorem is used. From reciprocity, < slot, dipole > = < dipole, slot >, so that

$$\iint E_x^s\big|_{z=h} I_x \, dx \, dy = \iint E_x^s\big|_{z=0} H_y^d\big|_{z=0} \, dx \, dy$$

The term Int2 therefore becomes

$$\text{Int2} = \frac{1}{V_s I_d} \iint E_x^s\big|_{z=0} H_y^d\big|_{z=0} \, dx \, dy$$

From Parseval's relation, this integral can be expressed as

$$\text{Int2} = \frac{4\pi^2}{V_s I_d} \int_{-\infty}^{\infty}\int_{-\infty}^{\infty} \tilde{H}_y^d\big|_{z=0} \tilde{E}_x^s\big|_{z=0} \, dk_x \, dk_y \tag{5.42}$$

To determine $\tilde{E}_x^s$ at $z = 0$, field distribution on the slot of length l is assumed as

$$E_x^s(y) = \frac{V_s}{s} \cos\left(\frac{\pi y}{l}\right) \hat{x}$$

which yields

$$\tilde{E}_x^s = \frac{V_s}{2\pi l} \frac{\cos(k_y l/2)}{[(\pi/l)^2 - k_y^2]} \frac{\sin(k_x s/2)}{(k_x s/2)}$$

Using the expressions for Int1 and Int2 from (5.41) and (5.42) in (5.40) we obtain I_d in terms of V_s.

5.3.3 Input Impedance of the Dipole

The input impedance of the dipole seen by the slot is obtained from the reaction of the dipole given to the slot. The input admittance is

$$Y_{\text{in}} = \frac{\langle \text{dipole, slot} \rangle}{V_s^2} = \frac{V_s I_d \text{Int2}}{V_s^2}$$

Eliminating V_s and I_d, we obtain

$$Y_{\text{in}} = -\frac{(\text{Int2})^2}{\text{Int1}} \tag{5.43a}$$

The input impedance Z_{in} is

$$Z_{\text{in}} = \frac{-\text{Int1}}{(\text{Int2})^2} \tag{5.43b}$$

Equation (5.43b) was computed numerically to obtain the input impedance of a dipole seen by a ground plane slot. The impedance is plotted in Figure 5.7 against the length of the dipole. For a large substrate thickness, the impedance is very high, implying a poor coupling between the slot and dipole. The coupling becomes tighter for a thinner substrate. It is worth mentioning here that, to obtain the total impedance, the self-impedance of the slot should also be added in parallel with that offered by the dipole alone.

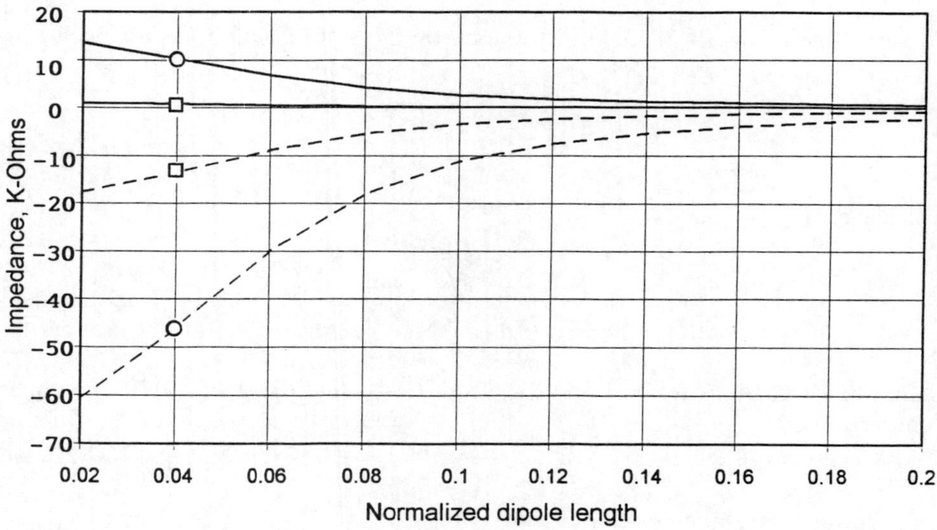

Figure 5.7 Dipole impedance versus normalized dipole length (a/λ_0): dipole width = slot width = $0.1\lambda_0$, $\varepsilon_r = 1.1$, slot length = $0.3\lambda_0$; —○— R ($h = 0.1\lambda_0$), —□— R ($h = 0.05\ \lambda_0$), --○-- X ($h = 0.1\lambda_0$), --□-- X ($h = 0.05\lambda_0$).

As mentioned before, the present formulation is a one-term approximation with regard to the dipole current distribution. This is done only to demonstrate the application of the spectral domain method for microstrip antenna analysis. For more accurate results, use a number of terms for the current distribution and then use the moment method technique to find the unknown currents. The procedure for finding the electromagnetic fields produced by each current function (in the moment method formulation, these are called *basis functions*) is same as detailed in Section 5.3.1. Extensive spectral moment method analysis of patches have been reported by several researchers.[4]

5.4 MICROSTRIP ARRAY ANTENNAS

The analysis of a finite-elements microstrip array antenna is not very straightforward, as in the case of an isolated element discussed in Sections 5.2 and 5.3. This is due to the mutual coupling effect between the elements in an array. In the edge admittance model, the mutual coupling effect can be incorporated via the active wall admittance of an element. The active wall admittance depends on the excitation coefficients of the neighboring elements. For details, the reader may refer to the published paper.[5]

To incorporate the mutual coupling effect in the patch current model, we need to estimate the current over the patch surface induced by the neighboring elements. This is a complicated task, beyond the scope of the book. For details of this procedure, the reader is referred to Pozar's work.[6]

For a uniformly excited (in magnitude) infinite array, the coupling effect can be included much more easily than a finite array. Because the elements are periodic in nature, a rectangular waveguide simulator with fewer elements can simulate the infinite array environment with regard to the impedance of an element. For a boresight array, where the elements are uniformly excited both in amplitude and phase, the array is equivalent to a single element inside a parallel plate waveguide as shown in Figure 5.8. This is due to the image effect caused by the electric and magnetic walls. The waveguide orientation depends on the polarization of the surface current on a patch element. A modal analysis of the waveguide structure would predict the input impedance of an element in the infinite array environment.

For an off-boresight infinite array, the simulator waveguide consists of a number of elements. For example, with reference to Figure 5.8 suppose the look angle

---

[4]For example, T. Itoh and W. Menzel, "A Full Wave Analysis Method of Open Microstrip Structures," *IEEE Trans. on Antennas and Propagation*, Vol. AP-29, 1981, pp. 63–68; D. M. Pozar, "A Reciprocity Method of Analysis for Printed Slot and Slot-Coupled Microstrip Antenna," *IEEE Trans. on Antennas and Propagation*, Vol. AP-34, 1986, pp. 1439–1446.

[5]A. K. Bhattacharyya and L. Shafai, "Effect of Mutual Coupling on the Radiation Pattern of Microstrip Phased Array Antenna," *IEEE AP-S International Symp. Digest*, 1986, pp. 891–893.

[6]D. M. Pozar, "Input Impedance and Mutual Coupling of Rectangular Microstrip Antennas," *IEEE Trans. on Antennas and Propagation*, Vol. AP-30, 1982, pp. 1191–1196.

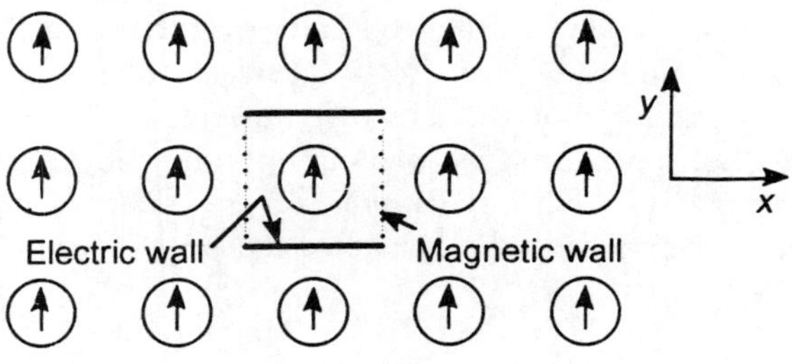

Figure 5.8 An infinite array for a boresight beam and its waveguide simulator.

is $\theta = 19.5°$ from the z-axis in the x-z plane. The phase difference between the adjacent elements (along x) will be $\Delta\phi$ ($= k_0 d \sin\theta$) $= 60°$ for an element spacing of $\lambda_0/2$. For a $-19.5°$ look angle (in the x-z plane), the phase distribution is reversed. Now, for symmetrical element shapes with symmetrical aperture fields, the input impedance of an element for both look angles will be identical. Therefore, the input impedance of an element will remain unchanged for an array that radiates in both $-19.5°$ and $+19.5°$ angles. The excitation coefficients of the elements for such an array can be found from the superposition of individual excitation coefficients. In this case the relative excitation coefficients will be 2, 2 cos60, 2 cos120, −2, −2 cos60, ..., and so on. This particular amplitude distribution can be simulated by a rectangular waveguide with only three elements, as shown in Figure 5.9. The input impedance of an element inside the waveguide and in an infinite array of 19.5° look angle will be identical. Therefore, the modal analysis will incorporate the mutual coupling effect in an infinite phased array. Extensive work on waveguide simulation of patch antennas are reported by Pozar and his group.[7]

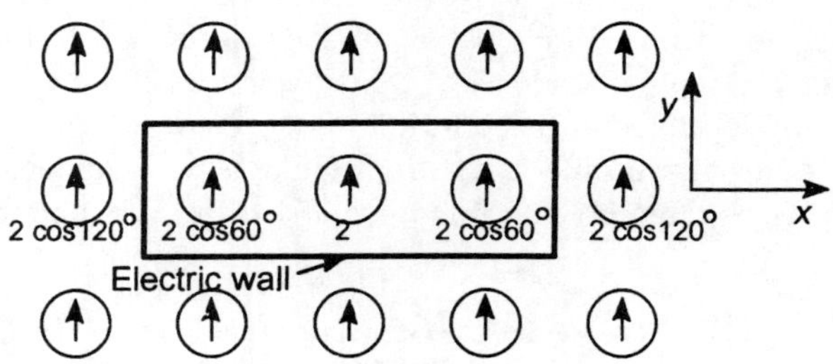

Figure 5.9 An infinite phased-array antenna with its waveguide simulator. The element spring is $\lambda_0/2$.

[7]For example, D. M. Pozar and D. H. Schaubert, "Scan Blindness in Infinite Phased Arrays of Printed Dipoles," *IEEE Trans. on Antennas and Propagation*, Vol. AP-32, 1984, pp. 602–610.

5.4.1 Scan Blindness

Scan blindness is an interesting (usually undesirable) phenomenon that occurs in microstrip phased-array antennas. For certain look angles, a microstrip phased-array antenna radiates no power. In this section, we will explain the phenomenon physically. A quantitative analysis will be presented later.

From Chapters 3 and 4 we have seen that a microstrip radiator always supports at least one surface wave mode that does not contribute to the radiated fields. The surface wave propagates along the air-dielectric interface. In a phased array, the desired look angle dictates the phase progression among the elements. If, for a particular look angle, the phase distribution among the elements is such that the element phase and the surface wave phase for all element locations differ by an integral multiple of 2π, the surface wave will be excited very strongly. The field intensity at all source locations will be very high. Therefore the active impedance of an element will be either very high or very low,[8] resulting in an impedance mismatch problem so that the element cannot radiate at all.

This can be further explained from the equivalent circuit of an array. Figure 5.10 depicts an array of magnetic line current sources. Figure 5.11 shows an equivalent circuit representation of the array.[9] The magnetic currents are y-directed and extended to infinity. The surface wave in this case will propagate along x direction.

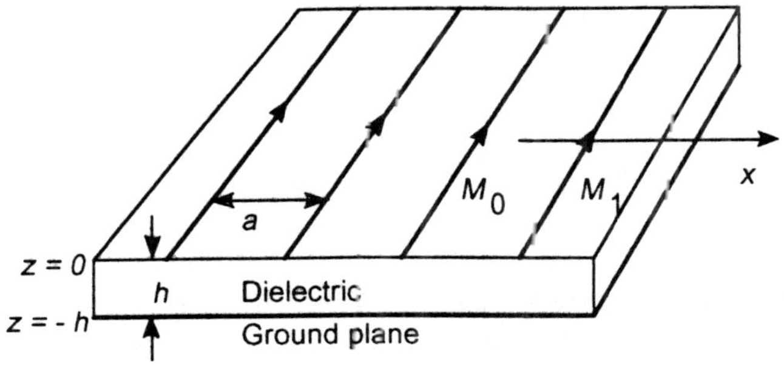

Figure 5.10 An array of magnetic line current sources over a dielectric-coated ground plane.

[8]This depends on the type of the radiating source.

[9]This equivalent circuit is an extension of the circuit shown in A. K. Bhattacharyya, "Effects of Ground Plane and Dielectric Truncation on the Efficiency of a Printed Structure," *IEEE Trans. on Antennas and Propagation*, Vol. AP-39, 1991, pp. 303–308.

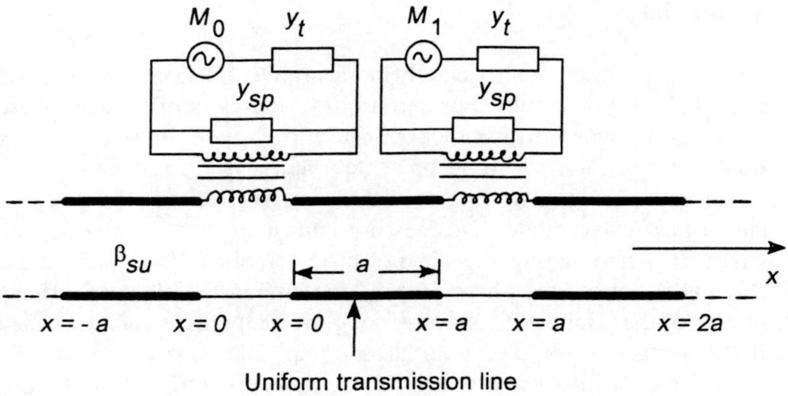

Figure 5.11 An equivalent circuit of the array in Figure 5.10.

In Figure 5.11, the uniform transmission line represents an unattenuated surface wave propagation. y_t is the equivalent Thevnin's admittance of the source. Here, y_{sp} is the space wave admittance, and β_{su} is the surface wave propagation constant. For a look angle θ, measured from the boresight direction, the phase difference between two adjacent elements should be $\Delta\phi = k_0 a \sin\theta$, where a is the distance between two adjacent elements, and k_0 is the wave number in free space. The surface wave phase difference between two successive elements locations is $\Delta\phi_{su} = \beta_{su} a$. The surface wave currents will be added in phase if

$$\beta_{su} a = k_0 a \sin\theta + 2n\pi \qquad n = 1, 2, 3, 4, \ldots \tag{5.44}$$

Note that n cannot be zero or negative because the surface wave number is always greater than free-space wave number. If (5.44) is satisfied, the amount of total current in the surface wave transmission line will be infinitely large due to *in-phase* addition of line currents from all sources. This will make the reflected impedance seen by the source very low. Because y_t is finite, the low reflected impedance causes an impedance mismatch. The available voltage across y_{sp} becomes negligibly small and the radiated power is insignificant.

Equation (5.44) gives the blind angle as a function of the parameters of the structure. The blind angle, θ, is given by

$$\theta = \sin^{-1}\left(\frac{\beta_{su}}{k_0} - \frac{2n\pi}{k_0 a}\right) = \sin^{-1}\left(\frac{\beta_{su}}{k_0} - \frac{n\lambda_0}{a}\right) \tag{5.45}$$

where λ_0 is the wavelength in free space. Figure 5.12 shows the plot of blind angle versus the substrate thickness. We considered $n = 1$ in this case, because that is the most probable case for the scan blindness to occur in a practical array. Note that the blind angle moves toward the boresight direction as the substrate thickness increases.

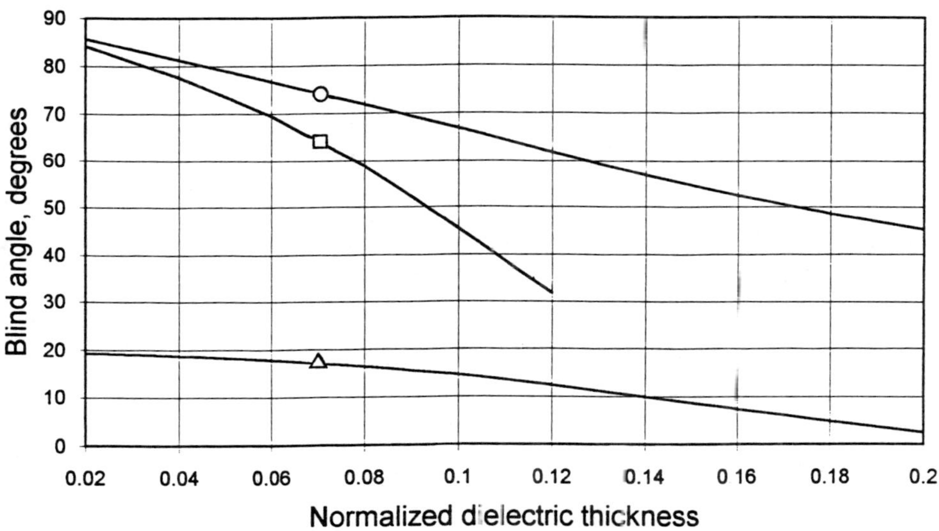

Figure 5.12 Blind angle versus normalized substrate thickness (h/λ_0): —○— $\varepsilon_r = 2.5$ and $a = 0.5\lambda_0$, —□— $\varepsilon_r = 5.0$ and $a = 0.5\ \lambda_0$, —△— $\varepsilon_r = 2.5$ and $a = 0.75\ \lambda_0$.

5.5 MATHEMATICAL ANALYSIS OF AN ARRAY OF SOURCES

This section presents a quantitative analysis of an infinite array of radiating sources on dielectric-coated ground plane. For the reason of simplicity, infinitely long magnetic currents are considered as the radiating sources (Figure 5.10). This makes a one-dimensional array, so that we can avoid the complexity of the two-dimensional problem.

The magnetic line currents are y directed, located at the air-dielectric interface of a dielectric-coated ground plane. We assume that the phase difference between two adjacent elements is ϕ_0. We will first calculate the fields produced by the array elements.

5.5.1. Fields Produced by the Array

The array source function can be represented by the following surface current lying on the surface $z = 0$:

$$M_y = M_0 \sum_{n=-\infty}^{\infty} \delta(x - na) \exp(-jn\phi_0) \tag{5.46}$$

Using the property of the Dirac-delta function, this equation is expressed as

$$M_y = M_0 \exp\left(-j\phi_0 \frac{x}{a}\right) \sum_{n=-\infty}^{\infty} \delta(x - na) \tag{5.47}$$

Now, the term $\Sigma_{n=-\infty}^{\infty} \delta(x - na)$ is a periodic function of x with periodicity a. We can expand the function in Fourier series. This yields

$$\sum_{n=-\infty}^{\infty} \delta(x - na) = \sum_{m=-\infty}^{\infty} a_m \exp(2\pi jmx/a)$$

where the Fourier coefficient a_m is given by

$$a_m = \frac{1}{a} \int_0^a \delta(x) \exp(-2\pi mjx/a)\, dx = \frac{1}{a} \tag{5.48}$$

Substituting (5.48) in (5.47), we obtain the following expression for M_y:

$$M_y = \frac{M_0}{a} \sum_{n=-\infty}^{\infty} \exp\{jx(2\pi n - \phi_0)/a\} \tag{5.49}$$

Because the magnetic current source is uniform along y, only the TM_z mode is produced. The spectral domain equivalent circuit will be similar to that shown in Figure 5.2. In the equivalent circuit, the source current I_0, is given by (see Table 2.1) $I_0 = j(\mathbf{k}_\rho \times \tilde{\mathbf{J}}_{sm}) \cdot \hat{z}$. For the present problem, $\tilde{\mathbf{J}}_{sm} = \hat{y}\tilde{M}_y$ where $\tilde{M}_y$ is the Fourier transform of M_y, which is

$$\tilde{M}_y = \frac{1}{2\pi} \int_{-\infty}^{\infty} M_y \exp(jk_x x)\, dx$$

Substituting the expression for M_y from (5.49) and using the following relation

$$\frac{1}{2\pi}\int_{-\infty}^{\infty} \exp(jtx)\,dx = \delta(t)$$

we obtain

$$\tilde{M}_y = \frac{M_0}{a}\sum_{n=-\infty}^{\infty} \delta\{k_x + (2n\pi - \phi_0)/a\}$$

In the equivalent circuit, the source current I_0 becomes

$$I_0 = jk_x\tilde{M}_y = jk_x\frac{M_0}{a}\sum_{n=-\infty}^{\infty} \delta\{k_x + (2n\pi - \phi_0)/a\} \tag{5.50}$$

The z component of the electric displacement vector is given by

$$D_z(x, y, z) = \int \frac{I_0}{y_{\text{in}}^m(k_x)} \exp(-jk_x x - jk_z z)\,dk_x$$

with

$$y_{\text{in}}^m(k_x) = \frac{j}{\varepsilon_0}\sqrt{k_0^2 - k_x^2} - \frac{1}{\varepsilon_0\varepsilon_r}\sqrt{k_1^2 - k_x^2}\tan\{h\sqrt{k_1^2 - k_x^2}\}$$

Substituting the expression for I_0 and performing the integration, we derive

$$D_z\,(x, y, z) = \frac{M_0}{a}\sum_n \frac{j\psi_n \exp(-j\psi_n x - jk_{zn} z)}{y_{\text{in}}^m(k_x = \psi_n)} \tag{5.51}$$

where $\psi_n = (\phi_0 - 2n\pi)/a$, $k_{zn} = \sqrt{k_0^2 - \psi_n^2}$. Equation (5.51) yields the total field produced by the array sources.

5.5.2 Grating Lobes

Equation (5.51) indicates that the array sources produce an infinite number of plane waves that propagate in various directions. The nth radiated plane wave corresponds to the nth term in the series, which is

$$D_{zn}(x, y, z) = \frac{M_0}{a} \frac{j\psi_n}{y_{\text{in}}^m(k_x = \psi_n)} \exp(-j\psi_n x - jk_{zn}z) \tag{5.52}$$

Let θ_n be the direction of propagation. The equation of the equiphase plane is given by $\psi_n x + k_{zn} z = 0$. Because $\psi_n^2 + k_{zn}^2 = k_0^2$, $(\psi_n/k_0, k_{zn}/k_0)$ represents the direction cosine of the normal of the equiphase plane. The normal makes an angle θ_n with the z-axis. Therefore, $\psi_n = k_0 \sin \theta_n$.

Suppose that the desired look angle is θ. Then $\phi_0 = k_0 a \sin \theta$. Therefore, $(\phi_0 - 2n\pi)/a = k_0 \sin \theta_n$, or $k_0 \sin \theta - (2n\pi/a) = k_0 \sin \theta_n$, which indicates that, when $n = 0$, the θ becomes equal to θ_n. That means that the plane wave associated with $n = 0$ propagates along the desired direction. The other plane waves for which $n \neq 0$ are the undesired grating lobes. The grating lobes exists in the visible region if θ_n is real, that is, $|\sin \theta_n| \leq 1$. The equality yields the optimum element spacing for which no grating lobe exists. If for all other values of n (except $n = 0$), the following condition holds

$$\left| \sin \theta - \frac{2n\pi}{k_0 a} \right| > 1 \qquad n = \pm 1, \pm 2, \ldots \tag{5.53}$$

then the array will produce no grating lobe in the visible region. This condition can be used to determine the spacing between two elements in an array. Suppose, we want to design a boresight array, so that $\theta = 0°$. For no grating lobe, $|2n\pi/k_0 a| > 1$, $n = \pm 1, \pm 2$, which yields $a \leq 2\pi/k_0 = \lambda_0$. The maximum element spacing for a boresight array is λ_0 (one free-space wavelength). Consider another example. Suppose, the desired look angle is $\theta = 90°$. For the nonexistence of any grating lobe, the condition is $|1 - 2n\pi/k_0 a| > 1$, $n = \pm 1, \pm 2, \ldots$. The condition is satisfied if $a \leq \lambda_0/2$.

5.5.3 Blind Angles

We have seen that if the element spacing is less than a half-wavelength, no grating lobe exists in the visible region, however, they will still exist in the invisible region (complex look angle). By choosing an optimum element spacing, we do not eliminate but rather move the grating lobe from the visible region to invisible region. In the case of a microstrip array the surface wave propagation can be treated as radiation in the invisible region. Radiation in the invisible region sometimes may cause impedance mismatch problem to the array elements. We will examine this situation in details. Let us focus on (5.51):

$$D_z = \frac{M_0}{a} \sum_n \frac{j\psi_n}{y_{\text{in}}^m(k_x = \psi_n)} \exp(-j\psi_n x - jk_{zn} z)$$

Now, y_{in}^m has a finite number of zeros (at least one). If it so happens that a zero of y_{in}^m coincides with a particular ψ_n, then D_z becomes infinitely large. This infinitely large value of D_z yields an infinitely large input admittance to the source creating a mismatch with the internal impedance of the source. The available voltage across y_{sp} in Figure 5.11 will be insignificant, so that the radiated power is negligibly small. The zero of y_{in}^m is a function of structure parameters and the operating frequency; ψ_n is a function of the desired look angle. For simplicity, assume that the structure supports only the lowest order surface wave mode for which $k_x = k_{x0}$. Let us assume that $\psi_n = k_{x0}$, that is, $(\phi_0 - 2n\pi)/a = k_{x0}$. Substituting the expression for ϕ_0, we have $(k_0 \sin\theta - 2n\pi)/a = k_{x0}$, yielding

$$\sin\theta = \frac{k_{x0}}{k_0} + \frac{2n\pi}{k_0 a} \qquad n = 0, \pm 1, \pm 2, \ldots$$

Now, θ is the scan angle, usually $0 < \theta < 90°$. For no grating lobe in the region $0 < \theta < 90°$, a must be less than $\lambda_0/2$. Also, k_{x0} is greater than k_0. The scan blind condition may occur when n is negative; that is,

$$\sin\theta = \left[\frac{k_{x0}}{k_0} + n\left(\frac{\lambda_0}{a}\right)\right] \tag{5.54}$$

A most likely case occurs when $n = -1$. In that case, the blind angle θ becomes

$$\theta = \sin^{-1}\left[\frac{k_{x0}}{k_0} - \frac{\lambda_0}{a}\right]$$

We obtained an identical expression for the blind angle from the equivalent circuit.

5.5.4. Impedance Analysis

The "scan blindness" effect would be very apparent from the impedance behavior of a source in an array environment. In this subsection, we determine the radiation impedance of a magnetic line current element. The radiation impedance is obtained from the complex radiated power. The complex power per unit length emanating from the nth element is

$$P_n = -M_0 \int \delta(x - na) \exp(-jn\phi_0) \cdot H_y^* |_{z=0} \, dx \tag{5.55}$$

Here, H_y is the total magnetic field component of the array. H_y can be calculated from D_z using the relation

$$\mathbf{H} = \frac{1}{\mu} \nabla \times \mathbf{A} \qquad \mathbf{A} = \hat{z} A_z$$

where

$$A_z = \frac{j\omega\mu}{a} M_0 \sum_n \frac{j \exp(-j\psi_n x - jk_{zn} z)}{\psi_n y_{\text{in}}^m(k_x = \psi_n)}$$

which is deduced from the relation $\tilde{A}_z = (j\omega\mu/k_\rho^2)\tilde{D}_z$. From this relation, H_y is derived as

$$H_y = \frac{-j\omega M_0}{a} \sum_n \frac{\exp(-j\psi_n x - jk_{zn} z)}{y_{\text{in}}^m(k_x = \psi_n)}$$

For simplicity, consider the element located at $x = 0$. The complex power becomes

$$P_0 = -|M_0|^2 \frac{j\omega}{a} \sum_n \frac{1}{[y_{\text{in}}^m(k_x = \psi_n)]^*}$$

Substituting the expression for y_{in}, one obtains

$$P_0 = |M_0|^2 \frac{j\omega\varepsilon_0}{a} \sum_n \frac{1}{-j\sqrt{k_0^2 - \psi_n^2} - (\sqrt{k_1^2 - \psi_n^2}/\varepsilon_r) \tan\{\sqrt{k_1^2 - \psi_n^2}\, h\}} \tag{5.56}$$

(Note: if $\psi_n > k_0$, then $\sqrt{k_0^2 - \psi_n^2}$ must be $j\sqrt{\psi_n^2 - k_0^2}$, the complex conjugate of $-j\sqrt{\psi_n^2 - k_0^2}$) with $\psi_n = k_0 \sin\theta - (2n\pi/a)$, θ being the scan angle. The radiation admittance per unit length would be

$$y_r = \frac{j\omega\varepsilon_0}{a} \sum_n \frac{1}{[j\sqrt{k_0^2 - \psi_n^2} + (\sqrt{k_1^2 - \psi_n^2}/\varepsilon_r) \tan\{\sqrt{k_1^2 - \psi_n^2}\, h\}]} \tag{5.57}$$

which is a divergent series of n. This is expected because a magnetic line current should have infinite susceptance. To resolve the nonconvergence problem, we assume that the magnetic current elements have finite thicknesses t along x. The Fourier transform of a source would contain the factor $\sin(k_x t/2)/(k_x t/2)$, so that the modified radiation admittance becomes

$$y_r = \frac{j\omega\varepsilon_0}{a}\sum_n \frac{\{\sin(\psi_n t/2)/(\psi_n t/2)\}^2}{[j\sqrt{k_0^2 - \psi_n^2} + (\sqrt{k_1^2 - \psi_n^2}/\varepsilon_r)\tan\{\sqrt{k_1^2 - \psi_n^2}h\}]} \tag{5.58}$$

This is a convergent series when $t \neq 0$. From the radiation admittance, we find the reflection coefficient

$$\Gamma = \frac{y_t - y_r}{y_t + y_r}$$

The internal admittance (y_t in Figure 5.11) is taken as the complex conjugate of the radiation admittance when the array radiates in the boresight direction. In other words, we assume that the array is matched for boresight radiation. Figure 5.13 shows the variation of the reflection coefficient (return loss) with the scan angle. As noted, for certain scan directions, the reflection coefficient approaches unity, implying no radiation situation.

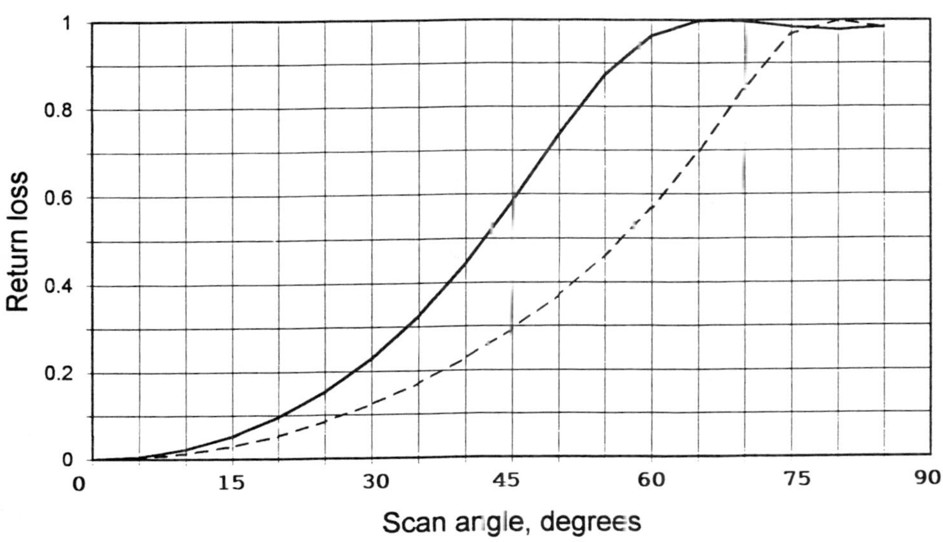

Figure 5.13 Reflection coefficient (return loss) versus scan angle of an infinite array of magnetic line current sources: $\varepsilon_r = 2.5$, —— $h = 0.1\ \lambda_0$, ---- $h = 0.05\ \lambda_0$.

5.6 SCREEN POLARIZER

The screen polarizer is an important application of a multilayered structure. A screen polarizer is used to convert a linearly polarized wave to a circularity polarized wave and vice versa. A typical screen polarizer consists of a dielectric slab with printed metal strips on both surfaces. The printed strips are usually uniform parallel lines. Other geometrical shaped strips have also been used to achieve a wideband performance.[10] In this section, we will confined ourselves to uniform parallel strips.

The mechanism of polarization conversion can be explained physically. Consider the polarizer screen shown in Figure 5.14. For the normal incidence of a plane wave, the electric field vector will have two components, one parallel to the strips and the other perpendicular to the strips. To the electric field component, which is parallel to the strips, the strip conductors work as inductors. This is primarily due to inductive stored energy generated by the strip currents. To the electric field component perpendicular to the strips, the conductors work as capacitive elements. The transverse electric fields accumulate line charges at the edges. The electric charges store electric energy. This results in capacitive effect to the y-polarized wave. The

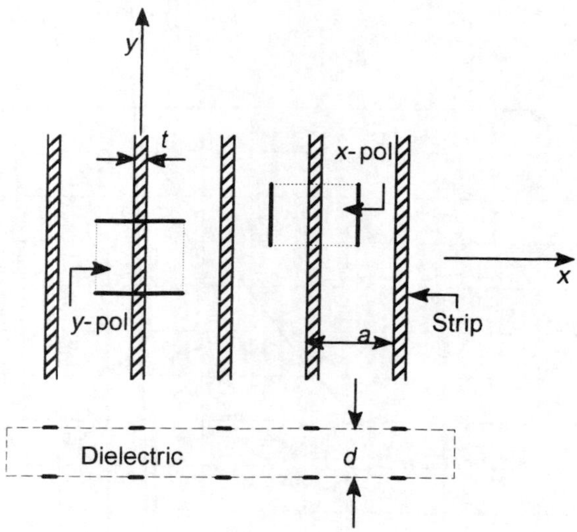

Figure 5.14 The structure of a printed strip polarizer. Waveguide simulators are also shown.

---

[10]R. S. Chu and K. M. Lee, "Analytical Model of a Multilayered Meander-Line Polarizer Plate with Normal and Oblique Plane-Wave Incidence," *IEEE Trans. on Antennas and Propagation*, Vol. 26 AP-35, 1987, pp. 652–661.

transmitted electric field components just after the first surface will have finite phase difference between them. The second surface will further add to this phase difference value, and the final transmitted field components may be in phase quadrature, which may yield a circularly polarized wave.

The analysis of a screen polarizer is very straightforward. For simplicity, we assume normal incidence. Also, we assume that the structure is infinitely large in both directions. Consider a y-polarized incident field. For y-polarized plane wave incidence, the structure is equivalent to a parallel plate waveguide with electric and magnetic walls, as shown in Figure 5.14. This is because the electric and magnetic conductors will not perturb the field distribution. The infinite structure problem is now reduced to a parallel plate waveguide problem. The waveguide has a metal strip discontinuity at the center, which can be characterized by an equivalent reactive element. There are two different media at two sides of the strip. To determine the reactance accurately, we need to know the unknown current that flows on the strip. In the following subsection, the unknown current is obtained.

5.6.1 y-Polarized Incident Field

For y-polarized plane wave incidence, the electric field vector is parallel to the strip lines. Therefore, the induced current on the strips will be uniform along y. The incident plane wave fields can be expressed as

$$E_y^{\text{inc}} = A \exp(-jk_0 z) \tag{5.59a}$$

$$H_x^{\text{inc}} = -\frac{A}{\eta} \exp(-jk_0 z) \tag{5.59b}$$

with $\eta = 120\ \pi$ ohms. For a narrow strip (which is usually the case) the current distribution is assumed to be

$$I_y = \frac{I_s}{\sqrt{1 - (2x/t)^2}} \tag{5.60}$$

The square root term in denominator accounts for the current singularity on the sharp edges. Our main objective is to relate the unknown current amplitude I_S with the known incident field amplitude A. To that end, we determine the electric fields produced by I_y and then impose the condition that the total tangential electric field (incident plus that produced by I_y) vanishes on the strip conductor. The condition yields I_y as a function of A. In the next step, we will determine the fields produced by I_y.

Fields Produced by Strip Current

The electric surface current I_y lies at the interface of two media. To find the fields, we employ the multilayer analysis. The current being uniform along y, the TM_z fields will not be excited. Only the TE_z mode fields exists. The TE_z fields are obtainable from B_z. The spectral domain equivalent circuit for the TE_z mode excited by an electric surface current is given in Table 2.1. Because the fields exist inside a wave-guide, the Fourier spectrums are discrete in nature, so that the Fourier transform is replaced by Fourier series. The source current I_0 in the TE_z equivalent circuit of Table 2.1 should be $I_0 = -j(\mathbf{k}_\rho \times \tilde{\mathbf{J}}_{se}) \cdot \hat{z}$. In the present case, $\mathbf{k}_\rho = (2m\pi/a)\,\hat{x}$, $m = 0, 1, 2, \ldots$, and $\tilde{\mathbf{J}}_{se} = \hat{y}\tilde{J}_{sy}$, where $\tilde{J}_{sy}$ is the Fourier coefficient of I_y as follows:

$$I_y = \sum_{m=-\infty}^{\infty} \tilde{J}_{sy} \exp\left(\frac{-j2m\pi x}{a}\right) \tag{5.61}$$

$$\tilde{J}_{sy} = \frac{1}{a}\int I_y \exp\left(\frac{j2m\pi x}{a}\right) dx \tag{5.62}$$

Substituting the expression for I_y, this integral simplifies to

$$\tilde{J}_{sy} = \left(\frac{\pi t}{2a}\right) I_s J_0\left(\frac{m\pi t}{a}\right) \tag{5.63}$$

where $J_0(m\pi t/a)$ represents the zero-order Bessel function. In the equivalent circuit, the source current I_0 therefore becomes

$$I_0 = -j\left(\frac{m\pi^2 t}{a^2}\right) I_s J_0\left(\frac{m\pi t}{a}\right) \tag{5.64}$$

The Fourier coefficient of the magnetic displacement vector component B_z is given by

$$\tilde{B}_z(z) = \frac{I_0}{y_{in}^e} \exp(-j\beta_2 z)$$

for $z > 0$, where

$$y_{in}^e = \frac{j\beta_1}{\mu_1} + \frac{j\beta_2}{\mu_2}$$

with $\beta_1 = \sqrt{k_1^2 - (2m\pi/a)^2}$, $\beta_2 = \sqrt{k_2^2 - (2m\pi/a)^2}$, and k_1, k_2 are the wave numbers in media 1 and 2, respectively. From Section 2.3.4, the electric field components can be written as

$$E_y = \omega \sum_{m=-\infty}^{\infty} \frac{1}{k_x} \frac{I_0}{y_{\text{in}}^e} \exp(-j\beta_2 z - jk_x x)$$

Substituting the expression for I_0 from (5.64) we obtain

$$E_y = -\frac{\omega\mu\pi t}{2a} I_s \sum_{m=-\infty}^{\infty} \frac{J_0(m\pi t/a)}{\beta_1 + \beta_2} \exp(-j\beta_2 z - jk_x x) \tag{5.65}$$

for $z > 0$. At $z = 0$, the electric field is of interest. The electric field component, at $z = 0$, becomes

$$E_y^{\text{strip}}(z = 0) = \frac{-\omega\mu\pi t}{a} I_s \sum_{m=0}^{\infty} \frac{J_0(m\pi t/a)}{(\beta_1 + \beta_2)\sigma_m} \cos\left(\frac{2m\pi x}{a}\right) \tag{5.66}$$

Relation Between the Strip Current and the Incident Field

To obtain I_s in terms of A, the boundary condition is imposed on the strip surface. The total field must vanish on the strip conductor at $z = 0$. Therefore,

$$E_y^{\text{strip}} + T_0 E_y^{\text{inc}} = 0 \qquad -t/2 < x < t/2$$

In this equation, T_0 is the transmission coefficient at the interface of two media, which is

$$T_0 = \frac{2\sqrt{\varepsilon_1}}{\sqrt{\varepsilon_1} + \sqrt{\varepsilon_2}}$$

We now multiply the above equation by $1/\sqrt{1 - (2x/t)^2}$ and integrate with respect to x over the strip conductor. This yields

$$T_0 A - \frac{\omega\mu\pi t}{a} I_s \sum_{m=0}^{\infty} \frac{[J_0(m\pi t/a)]^2}{\sigma_m(\beta_1 + \beta_2)} = 0$$

which gives

$$I_s = \frac{AT_0}{\left(\dfrac{\omega\mu\pi t}{a}\right)\displaystyle\sum_{m=0}^{\infty}\frac{J_0^2(m\pi t/a)}{\sigma_m(\beta_1+\beta_2)}} \tag{5.67}$$

5.6.2. Equivalent Admittance

The equivalent admittance is deduced from the TEM mode transmitted field. The total transmitted field is contributed by the incident field and the field produced by the induced current. The total TEM mode transmitted field at $z = 0+$ is $E_y^{\text{trns}} = T_0E_y^{\text{inc}} + E_y^{\text{strip}}$ (TEM mode only).

The TEM mode fields produced by the strip current is given by the first term ($m = 0$) in the series of (5.66). The TEM mode transmitted field becomes

$$E_y^{\text{trns}} = T_0A - \frac{\omega\mu\pi t I_s}{2ak_0(\sqrt{\varepsilon_1}+\sqrt{\varepsilon_2})} \tag{5.68}$$

Substituting the expression for I_s from (5.67), the transmission coefficient T is obtained as

$$T = \frac{E_y^{\text{trns}}}{A} = T_0 - \frac{T_0}{2k_0(\sqrt{\varepsilon_1}+\sqrt{\varepsilon_2})\displaystyle\sum_{m=0}^{\infty}\frac{J_0^2(m\pi t/a)}{\sigma_m(\beta_1+\beta_2)}} \tag{5.69}$$

The equivalent admittance offered by the strip can be calculated from the transmission coefficient T. Suppose y^{strip} is the equivalent admittance. The transmission coefficient T in terms of y^{strip} becomes

$$T = \frac{2\sqrt{\varepsilon_1}}{\sqrt{\varepsilon_1}+\sqrt{\varepsilon_2}+\eta y^{\text{strip}}}$$

which yields

$$y^{\text{strip}} = \frac{1}{\eta}\left(\frac{2\sqrt{\varepsilon_1}}{T} - \sqrt{\varepsilon_1} - \sqrt{\varepsilon_2}\right) \tag{5.70}$$

Substituting the expression for T from (5.69) and after simplification, the final expression for y^{strip} becomes

$$y^{\text{strip}} = \frac{1}{2\eta k_0 \sum_{m=1}^{\infty} \frac{J_0^2(m\pi t/a)}{(\beta_1 + \beta_2)}} \tag{5.71}$$

Recall that, β_1 and β_2 are the propagation constants for the TE_m modes in the two media. The distance between two adjacent strips, a, is usually less than a wavelength in the dielectric media so that β's are imaginary when $m > 0$. This makes y^{strip} imaginary. Therefore, for imaginary β_1 and β_2, the expression for the y^{strip} becomes

$$y^{\text{strip}} = \frac{-j}{2\eta k_0 \sum_{m=1}^{\infty} \frac{J_0^2(m\pi t/a)}{\left[\sqrt{(2m\pi/a)^2 - k_0^2\varepsilon_1} + \sqrt{(2m\pi/a)^2 - k_0^2\varepsilon_2}\right]}} \tag{5.72}$$

Note that the admittance is inductive. In Figure 5.15 we have plotted the susceptance versus the width of the strip with ε_2 as a parameter. Here, ε_1 is taken as unity. Expectedly, y^{strip} is found to increase with t.

In a similar manner, the equivalent strip admittance for a x-polarized incident field can be obtained. For the x-polarized field, the strip is equivalent to a capacitive

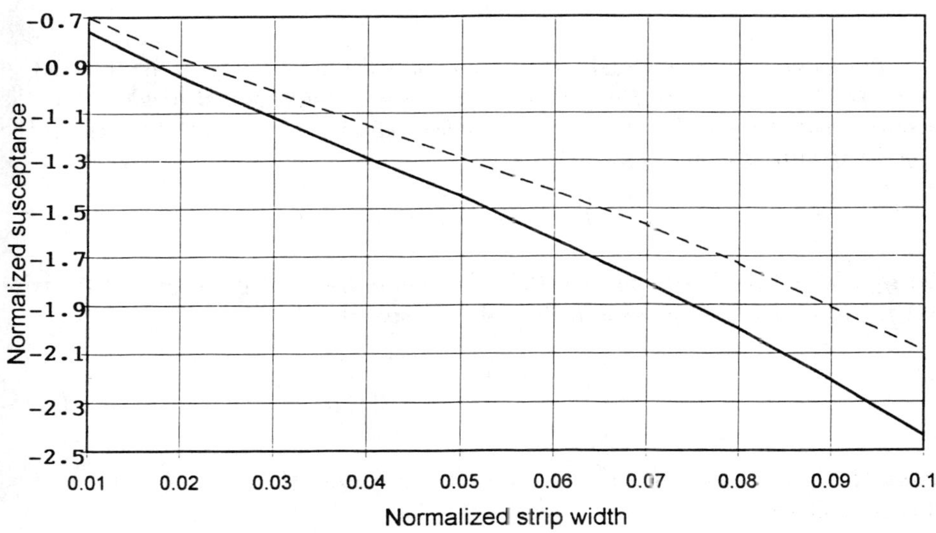

Figure 5.15 The equivalent susceptance of a strip versus strip normalized width (t/λ_0): $a = 0.4\ \lambda_0$: —— $\varepsilon_2 = 2.0$, ---- $\varepsilon_2 = 1.0$.

reactance. In a practical screen polarizer, the strips are usually very thin. The equivalent reactance is negligibly small because the strips do not reflect significant x-polarized power.

5.6.3 Analysis of a Two-Layered Polarizer

The circuit model can be used to analyze a multilayered screen polarizer. The equivalent circuit of a two-layered polarizer is shown in Figure 5.16. The shunt reactances represent the conducting strips. The transmission lines are the representatives of the dielectric media.

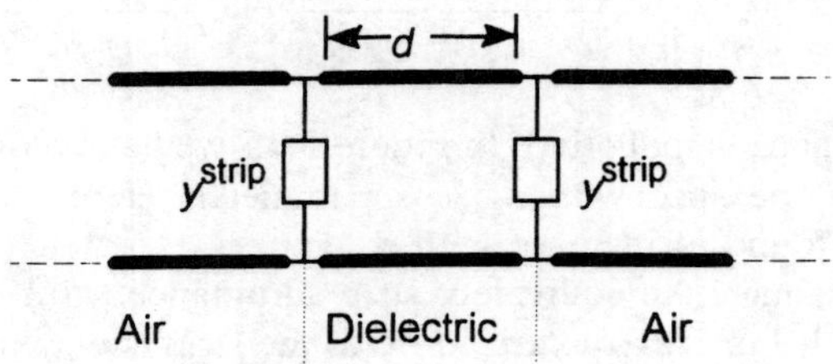

Figure 5.16 Equivalent circuit of a two-layered strip screen polarizer.

From the equivalent circuit, we can compute the axial ratio[11] of the transmitted wave and the return loss of an incident plane wave. Suppose the incident wave is linearly polarized and the electric field vector makes angle ψ with the x-axis. The x- and y-polarized components are

$$E_x = E_0 \cos \psi \qquad E_y = E_0 \sin \psi$$

The transmitted fields are obtained from the overall transmission matrix of the two-port network. The transmission matrix of the network is

$$[T] = \begin{bmatrix} T_{11} & T_{12} \\ T_{21} & T_{22} \end{bmatrix} = [T^{\text{strip}}][T^{\text{tl}}][T^{\text{strip}}] \tag{5.73}$$

where $[T^{\text{strip}}]$ represents the 2×2 transmission matrix of the shunt element y^{strip}. This is given by

[11]The axial ratio is defined as the ratio between the lengths of the minor axis and the major axis of the polarization ellipse. For a perfect circular polarized wave, the axial ratio is unity.

$$[T^{\text{strip}}] = \begin{bmatrix} 1 & 0 \\ y^{\text{strip}} & 1 \end{bmatrix} \tag{5.74}$$

The transmission matrix for the transmission line section is

$$[T^{\text{tl}}] = \begin{bmatrix} \cos \beta d & jz_0 \sin \beta d \\ jy_0 \sin \beta d & \cos \beta d \end{bmatrix} \tag{5.75}$$

where $\beta = k_0 \sqrt{\varepsilon_r}$ is the propagation constant of the transmission line, and $z_0 = \eta/\sqrt{\varepsilon_r}$, $y_0 = 1/z_0$. The transmission matrix $[T]$ relates the input voltage-current (V_i, I_i) to the output voltage-current (V_0, I_0). The output transmission line being matched, the output voltage and current are related to each other. The relation is $V_0 = I_0\eta$. The transmission matrix relation is

$$\begin{bmatrix} V_i \\ I_i \end{bmatrix} = [T] \begin{bmatrix} V_0 \\ I_0 \end{bmatrix} \tag{5.76}$$

The input reflection coefficient

$$\Gamma_i = \frac{\{(V_i/I_i) - \eta\}}{\{(V_i/I_i) + \eta\}} \tag{5.77}$$

can be determined in terms of the transmission parameters.

To find the axial ratio, the phase and amplitude of the output voltage V_0 must be known. Using (5.76), the output voltage V_0 can be expressed in terms of V_i. The input voltage, V_i, is related to the incident field and the relation is

$$V_i = (1 + \Gamma_i)\, E_y^{\text{inc}} \qquad (\text{or} \quad E_x^{\text{inc}}) \tag{5.78}$$

From (5.76) and (5.77), V_0 is expressed in terms of V_i. Then, using (5.78), V_0 is obtained in terms of E_x^{inc} and E_y^{inc}.

As mentioned before, for the x-polarized field, the value of y^{strip} is very close to 0, because the strip offers negligible susceptance to this polarization. The input reflection loss is determined by combining the x- and y-polarized reflected fields. The combined reflection coefficient is

$$\Gamma = \Gamma_{ix} \cos \psi + \Gamma_{iy} \sin \psi \tag{5.79}$$

where Γ_{ix} and Γ_{iy} are the reflection coefficients for the x- and y-polarized fields and should be obtained using (5.77). The output electric field vector is

$$\mathbf{E}_0 = \hat{x}V_{0x} + \hat{y}V_{0y} \tag{5.80}$$

where V_{0x} and V_{0y} are the output voltage for the x- and y-polarized incident fields, respectively. The strip being inductive for the y-polarized field, it is possible that the complex voltage V_{0y} is 90° out of phase with V_{0x} with equal magnitudes. This will give rise to a circular polarization. Furthermore, the separation between two strip planes may be adjusted to minimize the reflected power.

Figure 5.17 shows the variation of the input return loss and the output axial ratio of a typical screen polarizer. The incident fixed vector is at 45° angle with the x-axis. The polarizer performs well in a narrowband frequency. The bandwidth can be enhanced by adding more layers, which allows more gradual changes of phase values.

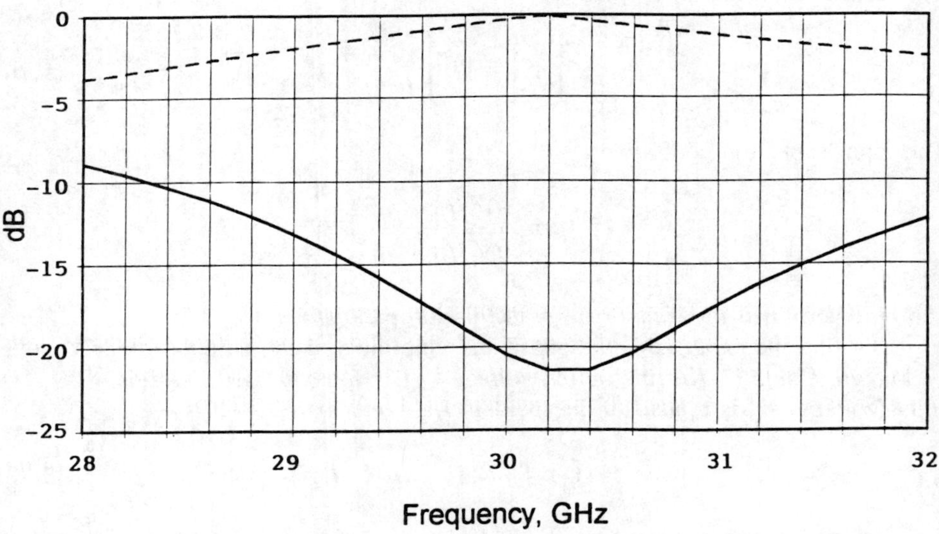

Figure 5.17 Input reflection loss and output axial ratio versus frequency of a typical two-layered strip screen polarizer: —— reflection loss, ----- axial ratio; a = 0.4 cm, t = 0.07 cm, d = 0.331 cm, ε_r = 1.2, incident polarization angle = 45° with the y axis.

PROBLEMS

5.1 A rectangular patch antenna of dimension $a \times b$ is excited by a coaxial feed. The feed is located at $x = a/2$ and $y = 0$. Use the magnetic wall assumption at $x = 0$ and $x = a$. The substrate thickness is h, and the dielectric constant is ε_r. Using the edge admittance model, obtain the expression for the input impedance seen by the probe.

5.2 The wall admittance of a rectangular patch antenna is defined as $y_w = -H_x/E_z$ for the wall at $y = 0$ and $y_w = H_x/E_z$ for the wall at $x = a$. For the TM_{10} mode operation, E_z is constant along a radiating aperture. Derive the expression for the wall admittance of the patch antenna.

5.3 From the wall admittance formulation, separate out the space wave and surface wave contributions. Obtain the radiation efficiency of the aperture.

5.4 A rectangular patch antenna is excited by a thin rectangular ground plane slot. The slot is symmetrically located at the center of the patch. The slot axis is parallel to the x-axis. The slot electric field has only a y component, which is given by $E_y^{\text{slot}} = E_0 \cos(\pi(x - a/2)/l)$, where $a \times b$ is the waveguide dimensions and l is the slot length. Replace the aperture by an equivalent magnetic surface current. For a very small slot width s, approximate the surface current as a line current. Then use the edge admittance model to find the impedance seen at the center of the slot. Use magnetic wall assumption at $x = 0$ and $x = a$.

5.5 The self-reaction of an electric surface current is defined as $\langle i, i \rangle = \int\int \mathbf{I}_s \cdot \mathbf{E}_s \, ds$, where $\mathbf{E}_s$ is the electric field vector produced by the electric surface current $\mathbf{I}_s$. Find the self-reaction of a uniform surface current element on a grounded dielectric structure. The current vector is $\mathbf{I}_s = \hat{x}I_0$, $-a/2 < x < a/2$, $-t/2 < y < t/2$, $z = h$. The ground plane is at $z = 0$, ε_r is the dielectric constant of the substrate, and h is the substrate thickness.

5.6 Assume that the structure in Problem 5.5 supports only the TM_0 surface wave mode. Obtain the radiation efficiency of the source.

5.7 Obtain the complex radiated power by a slot on a dielectric-coated ground plane. The structure supports the lowest order surface wave mode. Derive an expression for the surface wave power. Assume sinusoidal aperture field distribution on the slot. The slot length is l and width is s. The dielectric constant of the substrate is ε_r, and substrate thickness is h.

5.8 Find the blind angle of a one-dimensional microstrip scanned array antenna. The distance between two adjacent elements is $\lambda_0/2$, and the dielectric thickness $h = \lambda_0/20$, λ_0 is the free-space wavelength. The dielectric constant of the substrate is 2.5.

5.9 A microstrip phased array antenna consists of patches etched at the interface of two dielectric layers. The dielectric layer between the patch and the ground plane (substrate) has a dielectric constant of 2.0. The thickness of this layer is 0.159 cm. The dielectric material above the patch is also 0.159 cm. thick with a dielectric constant of 2.5. The array is designed for 5 GHz frequency. Compute the blind angle of the array. The element spacing is $\lambda_0/2$.

5.10 A microstrip phased array is designed to scan the beam from $\theta = 0°$ to $\theta = 75°$, where $\theta = 0°$ corresponds to the boresight direction. Determine the maximum possible element spacing so that no grating lobe occurs in the scanning region. What would be the maximum value of the substrate thickness so that the blind angle does not exist in the desired scanning range. The dielectric constant of the substrate is 2.25. The frequency of operation is 10 GHz.

5.11 A continuous magnetic surface current is lying over a dielectric-coated ground plane. The dielectric thickness is h, and the dielectric constant is ε_r. The current density is given by $\mathbf{M} = \hat{y}\exp(-jx\phi_0/a)$ $-\infty < x < \infty$

(a) Determine the radiation direction of the source.

(b) We define the radiation admittance per unit area as

$$y_r = \frac{1}{a}\int_0^a \mathbf{M} \cdot \mathbf{H}^* \, dx$$

where $\mathbf{H}$ is the magnetic field at the source location. Derive the expression for y_r as a function of ϕ_0.

(c) Assume $a = \lambda_0/2$, $h = \lambda_0/20$, $\varepsilon_r = 2.5$. Plot y_r versus ϕ_0. What can you conclude from this plot?

5.12 A screen polarizer consists of two identical strip screens separated by a dielectric layer of dielectric constant 1.2. The strip width is 0.04 cm, and the distance between the two adjacent strips is 0.4 cm. The thickness of the dielectric layer is 0.35 cm. The incident electric field vector is at 45° angle with the strips. Compute the input reflection coefficient and the output axial ratio for the operating frequency of 30 GHz.

Chapter 6
Analysis of Planar Transmission Lines

6.1 INTRODUCTION

This chapter presents the spectral domain analysis of planar transmission lines. They include the microstrip line, coupled lines, and the slot line. Spectral domain analysis is applicable for both static and time-harmonic problems. For electrically thin substrates, a planar transmission line supports the *quasi*-TEM mode, a mode that is not truly but very close to the TEM mode. The quasi-TEM mode analysis, which essentially is a static formulation, is presented for the microstrip line and coupled lines. The variational method is employed to find the capacitance of the lines. From the line capacitance, the effective dielectric constant and the characteristic impedance are obtained.

To obtain the dispersion characteristics of a transmission line, a full wave analysis is also presented for microstrip line and slot line structures. The full wave analysis assumes no particular supporting mode, therefore the method is very general. The dispersion behavior of both transmission lines are shown.

6.2 STATIC ANALYSIS OF MICROSTRIP LINE

The microstrip line is the simplest and most popular transmission line in planar circuit technology. It consists of a metal strip of finite width backed by a ground plane conductor (Figure 6.1). The region between the ground plane and the metal strip is filled with a dielectric material called *substrate*. Sometimes, the strip is also covered with another dielectric layer, called *superstrate*.

The propagating mode a microstrip line structure supports is very close to the TEM mode.[1] In this section we will assume that the propagating mode is TEM so

[1] In principle a microstrip line structure cannot support the TEM mode if the dielectric constant of the superstrate and substrate differ from unity. However, the characteristics of the fundamental supporting mode is very close to that of the TEM mode, and the TEM mode approximation works well, especially for the low-frequency range.

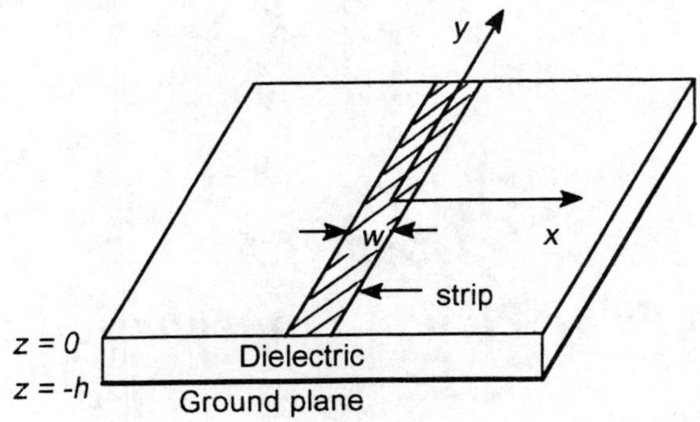

Figure 6.1 A microstrip line structure.

that the static formulation is applicable for the analysis. To obtain the propagation constant of the TEM mode, we need to determine the capacitance and inductance of the line per unit length. We will determine them in the following subsections.

6.2.1 Capacitance per Unit of Length

Suppose w is the width of the strip conductor, and h is the substrate thickness with ε_r as the dielectric constant. We assume an electrically thin strip, which is a valid approximation for many practical applications. For an accurate estimation of the capacitance, the variational technique is used. As a first step toward the determination of the static capacitance, we assume that the charge distribution on the strip is given by

$$\rho_{se} = \sum_{n=0}^{N-1} A_n(-1)^n \frac{T_{2n}(2x/w)}{\sqrt{1-(2x/w)^2}} \tag{6.1}$$

where $T_{2n}(\ )$, $n = 0, 1, 2, \ldots$ are the Tchebysheff polynomials[2] which satisfy the recurrence relation: $T_n(u) = 2uT_{n-1}(u) - T_{n-2}(u)$, with $T_0(u) = 1$, $T_1(u) = u$.

The advantage of using the Tchebysheff polynomials will be seen later. The factor in the denominator accounts for the edge conditions of the static charge distributions (see Problem 6.1). We will now use the spectral domain analysis to find the electric field produced by the charge distribution in (6.1). Because the charge is uniform along y, the electric field will be uniform along y. Also, the static field

[2]E. Kreyszig, *Advanced Engineering Mathematics*, 6th ed., New York, John Wiley, 1988, p. 234.

can be considered as the TM_z type because there is no magnetic field component[3] along z.

For the TM_z mode fields, we should obtain the electric displacement vector component, D_z, first. We express D_z in terms of the following Fourier integral:

$$D_z = \int_{-\infty}^{\infty} \tilde{D}_z (k_x, z) \exp(-jk_x x)\, dk_x \tag{6.2}$$

where $\tilde{D}_z$ is the Fourier transform of D_z. The general form of the spectral domain equivalent circuit for the TM_z mode was given in Table 2.1. The spectral domain equivalent circuit for the microstrip structure of Figure 6.1 is shown in Figure 6.2. The quantities y_{01}^m and y_{02}^m are the characteristic admittances of the transmission line sections. From Table 2.1, the expressions for y_{01}^m and y_{02}^m are

$$y_{01}^m = \frac{jk_{1z}}{\varepsilon_0 \varepsilon_r} \qquad y_{02}^m = \frac{jk_{0z}}{\varepsilon_0}$$

where $k_{1z} = \sqrt{k_1^2 - k_x^2}$, $k_{0z} = \sqrt{k_0^2 - k_x^2}$. Here k_1 and k_0 are the wave numbers in the dielectric region and free space, respectively. Because we are considering the static case, we have $k_1 = k_0 = 0$, so that $k_{0z} = k_{1z} = -j|k_x|$. The negative imaginary is taken to make the field decay along the positive z-direction. The characteristic admittances of the transmission lines therefore become

$$y_{01}^m = \frac{|k_x|}{\varepsilon_0 \varepsilon_r} \qquad y_{02}^m = \frac{|k_x|}{\varepsilon_0}$$

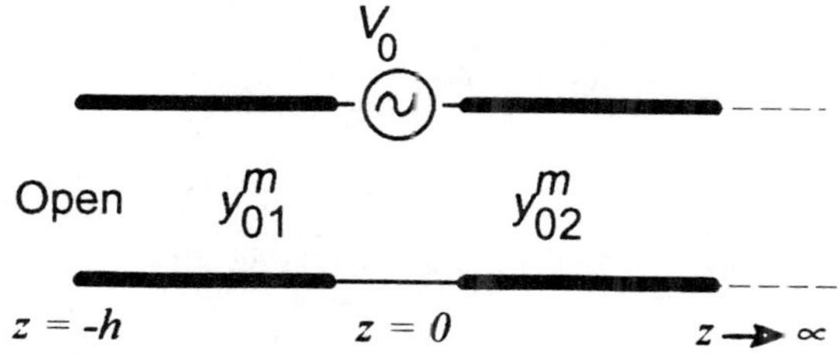

Figure 6.2 Spectral domain equivalent circuit of a microstrip line for the TM_z mode.

[3]For that matter, there is no magnetic field at any direction because no current exists on the conductor.

The source voltage V_0 is the Fourier spectrum of the surface charge density, which is given by

$$V_0 = \tilde{\rho}_{se} = \frac{1}{2\pi} \int \rho_{se} \exp(jk_x x)\, dx \tag{6.3}$$

Substituting the expression for ρ_{se} from (6.1) to (6.3), we obtain

$$V_0 = \frac{w}{4} \sum_{n=0}^{N-1} A_n J_{2n}(k_x w/2) \tag{6.4}$$

The line voltage in the equivalent circuit (Figure 6.2) at $z = 0-$ and $z = 0+$ are given by

$$V(z = 0-) = \frac{-V_0 z_1}{z_1 + z_2} \qquad V(z = 0+) = \frac{V_0 z_2}{z_1 + z_2} \tag{6.5}$$

where z_1 and z_2 are the equivalent impedances at the two sides of the voltage source V_0. They are given by

$$\begin{aligned} z_1 &= 1/jy_{01}^m \tan(k_{1z} h) \\ z_2 &= 1/y_{02}^m \end{aligned} \tag{6.6}$$

The voltage at $z = 0+$ is

$$V(z = 0+) = V_0 \frac{\tanh|k_x h|}{\varepsilon_r + \tanh|k_x h|}$$

The voltage distribution the region $z > 0$ is

$$V(z) = V(z = 0+) \exp(-jk_{0z} z) = \frac{V_0 \tanh|k_x h|}{\varepsilon_r + \tanh|k_x h|} \exp(-|k_x| z), \qquad z > 0$$

Recall that $V(z)$ is the Fourier spectrum of D_z. Therefore, in the space domain D_z would be

$$D_z = \int_{-\infty}^{\infty} \frac{V_0 \tanh|k_x h|}{\varepsilon_r + \tanh|k_x h|} \exp(-|k_x| z) \exp(-jk_x x)\, dk_x \tag{6.7}$$

The electrostatic potential Φ is given by

$$\partial\Phi/\partial z = -\frac{1}{\varepsilon_0} D_z$$

for $z > 0$, which yields

$$\Phi(x, z) = \frac{1}{\varepsilon_0} \int_{-\infty}^{\infty} \frac{V_0 \tanh|k_x h|}{|k_x|\ [\varepsilon_r + \tanh|k_x h|]} \exp(-|k_x|z) \exp(-jk_x x)\, dk_x \tag{6.8}$$

The integration constant is set to 0 to satisfy the boundary condition that $\Phi \to 0$ as $z \to \infty$. Substituting the expression for V_0 from (6.4) in (6.8), we obtain

$$\Phi = \frac{W}{4\varepsilon_0} \sum_{n=0}^{N-1} A_n \int_{-\infty}^{\infty} \frac{J_{2n}(k_x w/2) \tanh|k_x h|}{|k_x|[\varepsilon_r + \tanh k_x h|]} \exp[-jk_x x - |k_x|z]\, dk_x$$

The potential on the strip conductor is

$$\Phi(x, 0) = \frac{W}{4\varepsilon_0} \sum_{n=0}^{N-1} A_n \int_{-\infty}^{\infty} \frac{J_{2n}(k_x w/2) \tanh|k_x h|}{|k_x|[\varepsilon_r + \tanh|k_x h|]} \exp(-jk_x x)\, dk_x \tag{6.9}$$

The capacitance per unit length is obtained from the stationary expression. The stationary expression for the capacitance is[4]

$$C = \frac{Q^2}{\displaystyle\int_{-w/2}^{w/2} \Phi(x, 0)\, \rho_{se}\, dx} \tag{6.10}$$

where Q is the total charge per unit length on the strip. Q is given by

$$Q = \int_{-w/2}^{w/2} \rho_{se}\, dx$$

which, after substitution of ρ_{se} from (6.1), becomes

$$Q = \frac{\pi w}{2} A_0 \tag{6.11}$$

[4]R. E. Colin, *Field Theory of Guided Waves*, 2nd ed., New York, IEEE Press, 1991, Chapter 4.

Note that

$$\int_{-w/2}^{w/2} \frac{T_{2n}(2x/w)}{\sqrt{1-(2x/w)^2}}\,dx = 0$$

when $n \neq 0$.

Now, performing the integration appears in (6.10), we obtain

$$C = \frac{2\pi\varepsilon_0 A_0^2}{\displaystyle\int_{-\infty}^{\infty} \left[\sum_{n=0}^{N-1} A_n J_{2n}(k_x w/2)\right]^2 \frac{\tanh|k_x h|}{|k_x|[\varepsilon_r + \tanh|k_x h|]}\,dk_x}$$

The integrand in the denominator is an even function of k_x. Therefore, the expression for C reduces to

$$C = \frac{\pi\varepsilon_0 A_0^2}{\displaystyle\int_{0}^{\infty} \left[\sum_{n=0}^{N-1} A_n J_{2n}(k_x w/2)\right]^2 \frac{\tanh(k_x h)}{k_x[\varepsilon_r + \tanh k_x h]}\,dk_x} \tag{6.12}$$

Equation (6.12) yields the capacitance per unit length in terms of the unknown constants A_n. To determine A_n, we exploit the variational nature of the formulation. For simplicity, we assume $A_0 = 1$, which will not alter the final result.

Now, for the actual charge distribution, C is extremum. Any error in the charge distribution will move C away from its extreme value. Therefore, for the stationary condition, $\partial C/\partial A_i = 0$, $i = 1, 2, \ldots, N-1$, which is equivalent to

$$\frac{\partial}{\partial A_i}\int_{0}^{\infty} \left[\sum_{n=0}^{N-1} A_n J_{2n}\,(k_x w/2)\right]^2 \frac{\tanh(k_x h)}{k_x[\varepsilon_r + \tanh k_x h]}\,dk_x = 0$$

Upon simplification, this equation reduces to

$$\sum_{n=0}^{N-1} A_n B(i,n) = 0 \qquad i = 1, 2, \ldots, N-1 \tag{6.13}$$

with

$$B(i,n) = \int_{0}^{\infty} \frac{J_{2n}(k_x w/2) J_{2i}(k_x w/2)\tanh(k_x h)}{k_x[\varepsilon_r + \tanh k_x h]}\,dk_x \tag{6.14}$$

Solving (6.13) we obtain the values of A_n. We then substitute A_n in (6.12) to obtain the capacitance per unit length. It is worth mentioning that the spectral Galerkin's method and the variation formulation will yield an identical equation set for unknown constants A_n (see Problem 6.3). This is due primarily to the particular selection of the expansion functions in (6.1).

6.2.2 Effective Dielectric Constant

The effective dielectric constant is the ratio between the actual capacitance per unit of length and the capacitance per unit of length with the dielectric substrate replaced by air:

$$\varepsilon_{\text{eff}} = \frac{C(\varepsilon_r)}{C(\varepsilon_r = 1)} \tag{6.15}$$

Using the procedure detailed in the foregoing section, $C(\varepsilon_r)$ and $C(\varepsilon_r = 1)$ can be determined separately. One may verify from (6.12) that the capacitance C is an explicit function of two variables w/h and ε_r. This will be apparent if one substitutes $k_x w/2 = k'_x$ in the integral of (6.12).

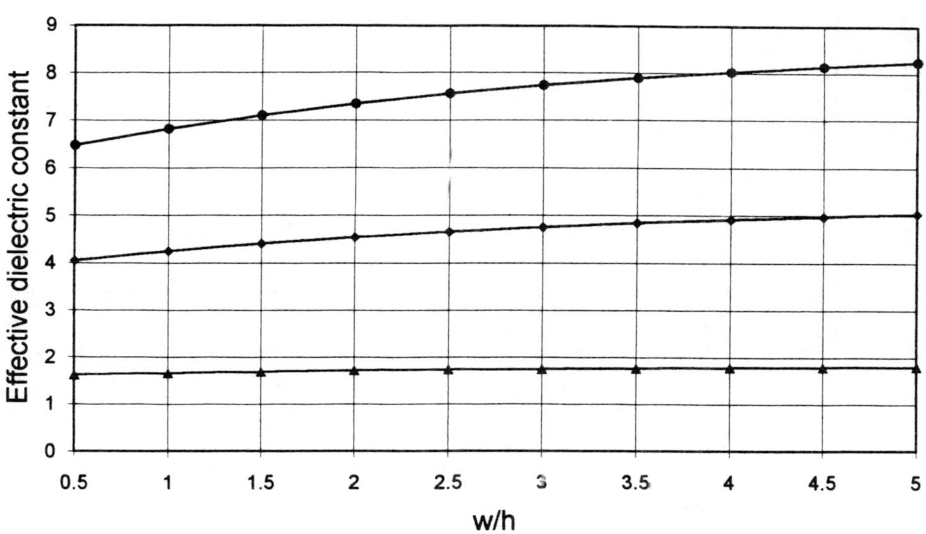

Figure 6.3 Effective dielectric constant of a microstrip line versus w/h: —▲▲— $\varepsilon_r = 2.0$, —◆◆— $\varepsilon_r = 6.0$, —●●— $\varepsilon_r = 10.0$

The solution for C is found to be rapidly convergent with regard to the number of expansion functions. Only three terms approximation ($N = 3$) yield less than 1% error for $\varepsilon_r \leq 4$ and $w/h \leq 5.0$. Figure 6.3 shows variation of ε_{eff} with w/h for various ε_r. The effective dielectric constant increases with w/h. These results agree very closely with the commonly used Wheeler's formula:[5]

$$\varepsilon_{\text{eff}} \approx \frac{\varepsilon_r + 1}{2} + \frac{\varepsilon_r - 1}{2}(1 + 10h/w)^{-1/2}$$

The maximum difference between the two results is 2%.

6.2.3 Characteristic Impedance

The characteristic impedance of a TEM transmission line is given by $Z_0 = \sqrt{L/C}$, where L is the inductance per unit of length and C is the capacitance per unit of length. The expression for C is derived in Section 6.2.1. The expression for the

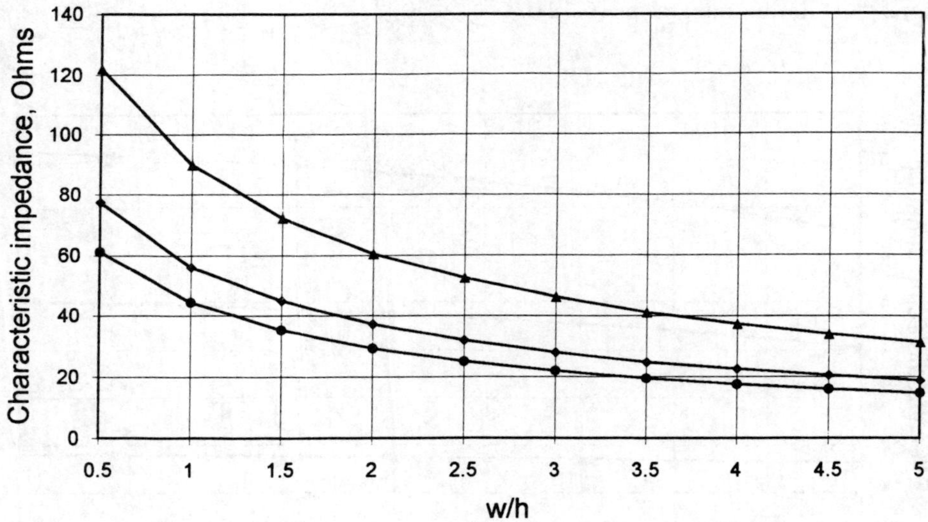

Figure 6.4 Characteristic impedance versus w/h of a microstrip line: —▲▲— $\varepsilon_r = 2.0$, —◆—◆— $\varepsilon_r = 6.0$, —●●— $\varepsilon_r = 10.0$.

[5]H. A. Wheeler, "Transmission Line Properties of Parallel Strips Separated by a Dielectric Sheet," *IEEE Trans. on Microwave Theory and Techniques*, Vol. MTT-13, 1965, pp. 172–185.

inductance L can be obtained from the capacitance C. It is well known from the transmission line theory that $\beta = k_0\sqrt{LC}$ is the propagation constant of the TEM mode. The propagation constant is also given by $\beta = k_0\sqrt{\mu_0\varepsilon_0\varepsilon_{\text{eff}}}$, which follows: $L = \mu_0\varepsilon_0\varepsilon_{\text{eff}}/C$. The characteristic impedance, Z_0, therefore, would be $Z_0 = \sqrt{L/C} = \sqrt{\mu_0\varepsilon_0\varepsilon_{\text{eff}}}/C$. Substituting the expression for ε_{eff} from (6.15), we deduce

$$Z_0 = \sqrt{\frac{\mu_0\varepsilon_0}{C(\varepsilon_r)\, C(\varepsilon_r = 1)}} \tag{6.16}$$

where the values of C are computed using (6.12) and (6.13). Figure 6.4 shows the variations of Z_0 with w/h. Expectedly, Z_0 decreases monotonically as w/h increases. It also decreases with ε_r.

6.3 COUPLED LINES ANALYSIS

Figure 6.5 shows the configuration of a coupled lines structure. It consists of two parallel microstrip lines. These types of transmission lines have applications in directional couplers, filters, and in several other planar microwave circuits.

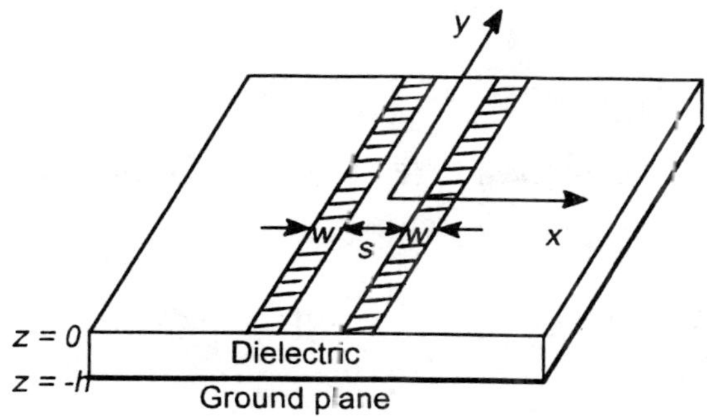

Figure 6.5 A microstrip coupled line.

The analysis of a coupled line is very similar to that of a microstrip line. From the circuit point of view, a coupled line can be completely characterized by four parameters: even-mode characteristic impedance, odd-mode characteristic impedance, even-mode effective dielectric constant, and odd-mode effective dielectric constant. The even mode is defined as the propagating mode for which the current in one strip conductor is even symmetric to the current on the other strip. For the odd mode, the strip currents are odd symmetric to each other. We will present the analyses for both modes.

6.3.1 Even-Mode Analysis

We have seen in Section 6.2.2 that, to obtain the effective dielectric constant, we need to determine the capacitance per unit of length. For the even mode, the charge distribution on one line will be the mirror image of that of the other line. The charge distribution can be expressed as

$$\rho_{se} = \sum_{n=0}^{N-1} \frac{A_n T_n \left\{ \dfrac{2(x - s/2 - w/2)}{w} \right\}}{\sqrt{1 - \left[\dfrac{2(x - s/2 - w/2)}{w} \right]^2}} \qquad s/2 < x < s/2 + w$$

$$= \sum_{n=0}^{N-1} \frac{A_n T_n \left\{ \dfrac{-2(x + s/2 + w/2)}{w} \right\}}{\sqrt{1 - \left[\dfrac{2(x + s/2 + w/2)}{w} \right]^2}} \qquad -s/2 - w < x < -s/2$$

$$= 0 \qquad \text{otherwise.}$$

The potential function Φ is given in (6.8). For the present problem the term V_0 in (6.8) would be

$$V_0 = \tilde{\rho}_{se} = \frac{1}{2\pi} \int_{-s/2-w}^{s/2+w} \rho_{se} \exp(jk_x x)\, dx$$

$$= \frac{w}{2} \sum_{n=0}^{N-1} A_n J_n(k_x w/2) \cos\left\{ k_x \left(\frac{s+w}{2} \right) - \frac{n\pi}{2} \right\}$$

Following the steps in Section 6.2.1, the capacitance per unit of length for each line is obtained as

$$C_{\text{even}} = \frac{\pi \varepsilon_0 A_0^2}{2 \displaystyle\int_0^{\infty} \left[\sum_{n=0}^{N-1} A_n J_n(k_x w/2) \cos\left\{ k_x \left(\frac{s+w}{2} \right) - \frac{n\pi}{2} \right\} \right]^2 \frac{\tanh(k_x h)}{k_x [\varepsilon_r + \tanh(k_x h)]}\, dk_x} \tag{6.17}$$

For C_{even} to be stationary, $\partial C_{\text{even}}/\partial A_i = 0$, which yields

$$\sum_{n=0}^{N-1} A_n B(i, n) = 0 \tag{6.18}$$

with

$$B(i, n) = \int_0^\infty \frac{J_n(k_x w/2) J_i(k_x w/2) \cos\left\{k_x \frac{s+w}{2} - \frac{n\pi}{2}\right\} \cos\left\{k_x \frac{s+w}{2} - \frac{i\pi}{2}\right\} \tanh(k_x h)}{k_x[\varepsilon_r + \tanh(k_x h)]} dk_x \tag{6.19}$$

Solving (6.18) we obtain values for A_n, which are then substituted in (6.17) to yield C_{even}. For the effective dielectric constant and the characteristic impedance, the formulas in (6.15) and (6.16) are used. To calculate C_{even} ($\varepsilon_r = 1$), the process is repeated for $\varepsilon_r = 1$. Equations (6.15) and (6.16) are then used to complete ε_{eff} and Z_0 for the even mode. Figures 6.6 and 6.7 show typical plots of ε_{eff} and Z_0 for even-mode propagation.

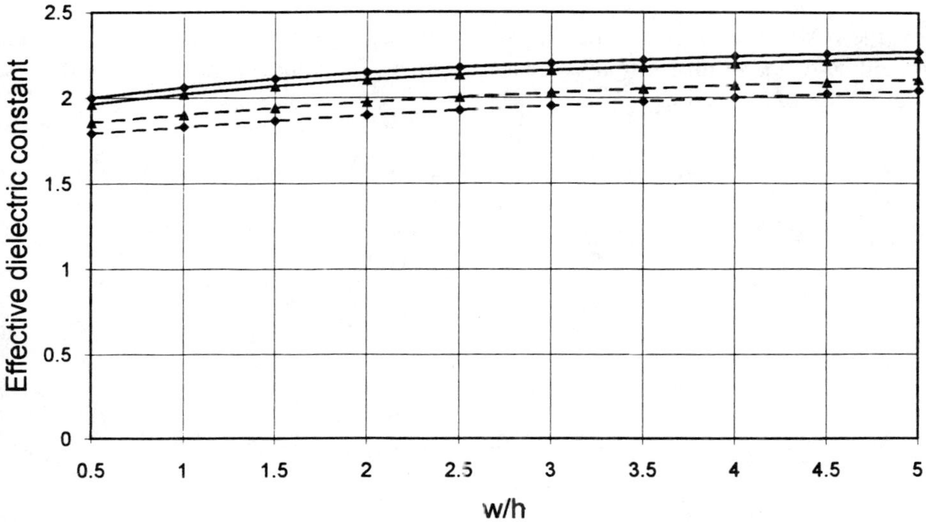

Figure 6.6 Effective dielectric constant versus w/h of a microstrip coupled line: $\varepsilon_r = 2.5$; —— even mode, --------- odd mode, ◆◆ $s/h = 1.0$, ▲▲ $s/h = 3.0$.

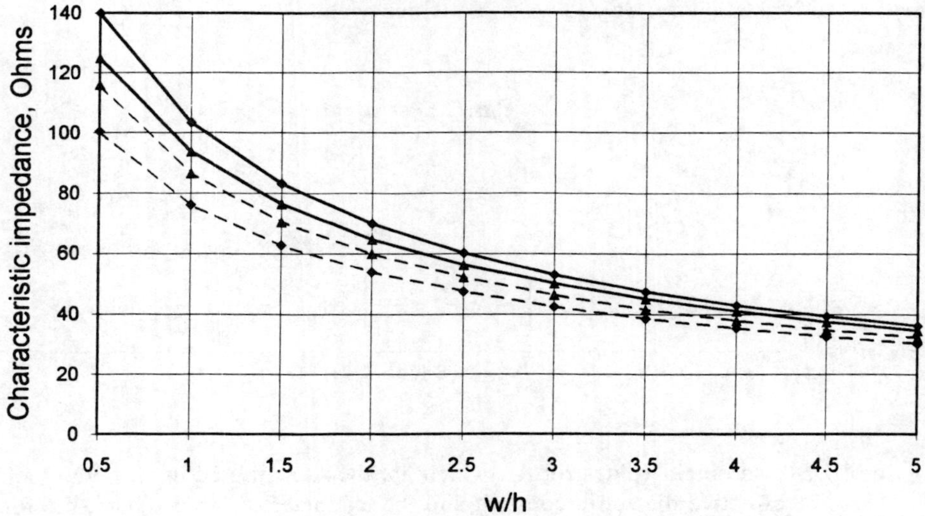

Figure 6.7 Variation of characteristic impedance with w/h of a coupled line: $\varepsilon_r = 2.5$; —— even mode, ------ odd mode, ◆◆ $s/h = 1.0$, ▲▲ $s/h = 3.0$.

6.3.2 Odd-Mode Analysis

For the odd mode analysis, the electric charge distributions on the strip conductors should be negative mirror images of one another. The charge distribution in this case may be approximated as

$$\rho_{se} = \sum_{n=0}^{N-1} A_n \frac{T_n\left\{\dfrac{2(x - s/2 - w/2)}{w}\right\}}{\sqrt{1 - \left[\dfrac{2(x - s/2 - w/2)}{w}\right]^2}} \qquad s/2 < x < s/2 + w$$

$$= -\sum_{n=0}^{N-1} A_n \frac{T_n\left\{\dfrac{-2(x + s/2 + w/2)}{w}\right\}}{\sqrt{1 - \left[\dfrac{2(x + s/2 + w/2)}{w}\right]^2}} \qquad -s/2 - w < x < -s/2$$

$$= 0 \qquad \text{otherwise}$$

The odd mode capacitance, C_{odd}, is derived as

$$C_{\text{odd}} = \frac{\pi \varepsilon_0 A_0^2}{2\int_0^\infty \left[\Sigma A_n J_n(k_x w/2) \sin\left\{k_x \frac{s+w}{2} - \frac{n\pi}{2}\right\}\right]^2 \frac{\tanh(k_x h)}{k_x[\varepsilon_r + \tanh(k_x h)]} dk_x} \tag{6.20}$$

where the sine term in the denominator is due to the Fourier integral of the odd function, ρ_{se}. The unknown constants are obtained through solving the following linear equations.

$$\sum_{n=0}^{N-1} A_n B(i, n) = 0 \qquad i = 0, 1, 2, \ldots N-1 \tag{6.21}$$

where $B(i, n)$ for the odd mode case is

$$B(i, n) = \int_0^\infty \frac{J_n(k_x w/2) J_i(k_x w/2) \sin\left\{k_x \frac{s+w}{2} - \frac{n\pi}{2}\right\} \sin\left\{k_x \frac{s+w}{2} - \frac{i\pi}{2}\right\} \tanh(k_x h)}{k_x[\varepsilon_r + \tanh(k_x h)]} dk_x \tag{6.22}$$

The effective dielectric constant, ε_{eff}, and characteristic impedance, Z_0, are obtainable from the values of C_{odd}, calculated for the actual substrate and the air substrate. In Figures 6.6 and 6.7 the odd-mode effective dielectric constant and characteristic impedance are plotted. For the same physical parameters, the odd-mode characteristic impedance is lower then the even-mode characteristic impedance. For the odd mode, the amount of fringing fields outside the dielectric region is larger then that of the even mode, due to the presence of opposite polarity electric charge near the edge. This effectively increases the value of capacitance per unit of length, resulting in a lower characteristic impedance. Furthermore, the odd and even-mode characteristics approach each other as the line separation, s, increases (see Figures 6.6 and 6.7).

6.4 FULL WAVE ANALYSIS OF THE MICROSTRIP LINE

Sections 6.2 and 6.3 present the static formulations for microstrip line and coupled lines. The static analysis provides fairly accurate result when the width of the strip and thickness of the substrate are small compared to the wavelength of operation. The static approximation becomes inaccurate at a higher frequency range. In this section, we present a full wave analysis of a microstrip line. As a first step, the conducting strip is replaced by unknown electric currents. The TE_z and TM_z field components produced by the strip currents are then obtained by using the spectral domain analysis developed in Chapter 2. The boundary condition that the tangential electric field components vanish on the strip conductor is then imposed. This yields a set of linear homogeneous equations with unknown current amplitudes. From the solvability criterion, the propagation constant is obtained. The characteristic impedance is obtained from the ratio of the voltage and current of a guided wave.

The geometry of a microstrip line is shown in Figure 6.1. The unknown surface current components on the strip are assumed as

$$I_x = Af(x)\exp(-j\beta y) = AT_1\left(\frac{2x}{w}\right)\sqrt{1-\left(\frac{2x}{w}\right)^2}\exp(-j\beta y) \tag{6.23}$$

$$I_y = Bg(x)\exp(-j\beta y) = \frac{B}{\sqrt{1-\left(\frac{2x}{w}\right)^2}}\exp(-j\beta y) \tag{6.24}$$

where $T_1(2x/w)$ is the first-order Tchebysheff polynomial, which is defined in Section 6.2.1. Such current distribution functions are chosen to satisfy the edge conditions. A and B are unknown coefficients; β is the propagation constant to be found. The reader may note that we are limiting calculations to one term approximation for each current component. For most the practical situations, this approximation is found to be sufficiently accurate when $w/h \leq 4.0$. However, one may use more then one term for I_x and I_y in (6.23) and (6.24) and proceed, using the procedure presented here. For more then one term, one has to use the Galerkin's method to find the unknowns.

6.4.1 The TM_z Electric Field

The TM_z mode field components are found from the z component of the electric displacement vector D_z. The Fourier spectrum of D_z is obtained from the spectral domain equivalent circuit in Figure 6.2. The characteristic admittance and the propagation constant of a transmission line section are given in Table 2.1. In the equivalent transmission line the source voltage V_0 is the two-dimensional Fourier transform of the electric surface charge density, $\tilde{\rho}_{se}$. The Fourier transform of the surface charge density $\tilde{\rho}_{se}$ in this case is

$$\tilde{\rho}_{se} = V_0 = \frac{1}{\omega}\left[Ak_xF(k_x) + B\beta G(k_x)\right]\delta(k_y - \beta) \tag{6.25}$$

where $AF(k_x)$ and $BG(k_x)$ are one-dimensional Fourier transforms of x and y current components, respectively. $F(k_x)$ and $G(k_x)$ are obtained as

$$F(k_x) = \frac{j}{2k_x} J_2(k_x w/2)$$

$$G(k_x) = \frac{w}{4} J_0(k_x w/2)$$

From the equivalent circuit, $\tilde{D}_z$ is obtained as

$$\tilde{D}_z = V_0 \left\{ \frac{j y_{01}^m \tan k_{1z} h}{y_{02}^m + j y_{01}^m \tan k_{1z} h} \right\} \exp(-jk_{0z} z) \qquad z > 0 \tag{6.26a}$$

$$= -V_0 \left\{ \frac{y_{02}^m}{y_{02}^m + j y_{01}^m \tan k_{1z} h} \right\} \frac{\cos k_{1z}(z+h)}{\cos k_{1z} h} \qquad -h < z < 0 \tag{6.26b}$$

Following Section 2.2.4, the TM$_z$ mode electric field components are (for $z > 0$)

$$E_x^m = -\frac{1}{\varepsilon_0} \int_{-\infty}^{\infty} \int_{-\infty}^{\infty} \frac{k_x k_{0z} \tilde{D}_z}{k_x^2 + \beta^2} \exp(-jk_x x - jk_y y)\, dk_x\, dk_y \tag{6.27}$$

$$E_y^m = -\frac{1}{\varepsilon_0} \int_{-\infty}^{\infty} \int_{-\infty}^{\infty} \frac{k_y k_{0z} \tilde{D}_z}{k_x^2 + \beta^2} \exp(-jk_x x - jk_y y)\, dk_x\, dk_y \tag{6.28}$$

Recall that $\tilde{D}_z$ contains a Dirac-delta function, namely, $\delta(k_y - \beta)$, so that these integrals reduce to one-dimensional integrals.

6.4.2 The TE$_z$ Electric Field

The TE$_z$ field components are obtainable from the z component of the magnetic displacement vector, B_z. The Fourier spectrum of B_z corresponds to the line voltage in the spectral domain equivalent circuit (see Table 2.1). The spectral domain equivalent circuit is shown in Figure 6.8. The source current I_0 is given in Table 2.1, which in the present situation becomes

$$I_0 = -j(\hat{x}k_x + \hat{y}k_y) \times (\hat{x}\tilde{I}_x + \hat{y}\tilde{I}_y)$$

$$= j[Ak_y F(k_x) - Bk_x G(k_x)]\delta(k_y - \beta) \tag{6.29}$$

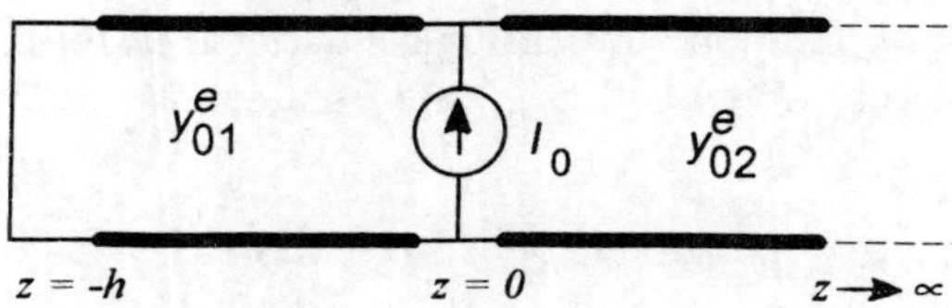

Figure 6.8 Spectral domain equivalent circuit for the TE_z mode of a microstrip line shown in Figure 6.1.

The line voltage $\tilde{B}_z$ is

$$\tilde{B}_z = I_0 \frac{\exp(-jk_{0z}z)}{[y^e_{02} - jy^e_{01}\cot(k_{1z}h)]}, \qquad z \geq 0 \tag{6.30}$$

with $y^e_{01} = jk_{1z}/\mu_0$, $y^e_{02} = jk_{0z}/\mu_0$.

Following the Section 2.3.4, the TE_z electric field components are obtained. The final expressions are

$$E^e_x = -\omega \int_{-\infty}^{\infty}\int_{-\infty}^{\infty} \frac{k_y \tilde{B}_z}{k_x^2 + \beta^2} \exp(-jk_x x - jk_y y)\, dk_x\, dk_y \tag{6.31}$$

$$E^e_y = \omega \int_{-\infty}^{\infty}\int_{-\infty}^{\infty} \frac{k_x \tilde{B}_z}{k_x^2 + \beta^2} \exp(-jk_x x - jk_y y)\, dk_x\, dk_y \tag{6.32}$$

Like E^m_x and E^m_y, these integrals are also reduced to single integrals because $\tilde{B}_z$ consists of the $\delta(k_y - \beta)$ factor.

6.4.3 The Propagation Constant

To find the unknown propagation constant β, the electric field boundary condition is imposed on the strip conductor. The tangential field components must vanish everywhere on the strip, which yields

$$E^m_x + E^e_x = 0 \qquad -w/2 < x < w/2 \qquad z = 0$$

$$E^m_y + E^e_y = 0 \qquad -w/2 < x < w/2 \qquad z = 0$$

We now use the weighted average values of the electric field components on the strip. Following Galerkin's method of moment, the weighing function (also known as the *test function*) is taken to be same as the current distribution function (in moment method terminology, *basis function*). This follows:

$$\langle E_x^m, f(x)\rangle + \langle E_z^e, f(x)\rangle = 0 \tag{6.33a}$$

$$\langle E_y^m, g(x)\rangle + \langle E_y^e, g(x)\rangle = 0 \tag{6.33b}$$

where $f(x)$ and $g(x)$ are given in (6.23) and (6.24), respectively. The symbol $\langle t_1(x), t_2(x)\rangle$ denotes the inner product of two functions $t_1(x)$ and $t_2(x)$, which is defined as

$$\langle t_1(x), t_2(x)\rangle = \int_x t_1(x)\, t_2(x)\, dz$$

Collecting the coefficients of A and B, (6.33) become

$$A[E_{xa}^m + E_{xa}^e] + B[E_{xb}^m + E_{xb}^e] = 0 \tag{6.34a}$$

$$A[E_{ya}^m + E_{ya}^e] + B[E_{yb}^m + E_{yb}^e] = 0 \tag{6.34b}$$

where

$$E_{xa}^m = \frac{-2\pi}{\omega}\int_{-\infty}^{\infty} \frac{k_x^2 k_{0z}}{\beta^2 + k_x^2} |F(k_x)|^2 R_m\, dk_x \tag{6.35a}$$

$$E_{xb}^m = \frac{-2\pi\beta}{\omega}\int_{-\infty}^{\infty} \frac{k_x k_{0z}}{\beta^2 + k_x^2} G(k_x)F^*(k_x)R_m\, dk_x \tag{6.35b}$$

$$E_{xa}^e = -2\pi j\omega\beta^2 \int_{-\infty}^{\infty} \frac{|F(k_x)|^2}{k_x^2 + \beta^2} R_e\, dk_x \tag{6.35c}$$

$$E_{xb}^e = 2\pi j\omega\beta \int_{-\infty}^{\infty} \frac{k_x}{k_x^2 + \beta^2} G(k_x)\, F^*(k_x)R_e\, dk_x \tag{6.35d}$$

$$E_{ya}^m = -E_{xb}^m$$

$$E_{yb}^m = \frac{-2\pi\beta^2}{\omega}\int_{-\infty}^{\infty} \frac{k_{0z}}{\beta^2 + k_x^2} |G(k_x)|^2 R_m\, dk_x \tag{6.35e}$$

$$E^e_{ya} = -E^e_{xb} \tag{6.35f}$$

$$E^m_{yb} = -2\pi j\omega \int_{-\infty}^{\infty} \frac{k_x^2}{k_x^2 + \beta^2} |G(k_x)|^2 R_e \, dk_x \tag{6.35g}$$

with

$$R_m = \frac{jy^m_{01} \tan k_{1z}h}{\varepsilon_0[y^m_{02} + jy^m_{01} \tan k_{1z}h]} \qquad R_e = \frac{j \tan k_{1z}h}{y^e_{01} + jy^e_{02} \tan k_{1z}h}$$

$$k_{1z} = \sqrt{k_1^2 - k_x^2 - \beta^2} \qquad k_{0z} = \sqrt{k_0^2 - k_x^2 - \beta^2}$$

Equations (6.34) have nontrivial solutions for A and B if

$$\begin{vmatrix} E^m_{xa} + E^e_{xa} & E^m_{xb} + E^e_{xb} \\ E^m_{ya} + E^e_{ya} & E^m_{yb} + E^e_{yb} \end{vmatrix} = 0 \tag{6.36}$$

Equation (6.36) is a transcendental equation of β. The integrals in (6.35) generally have a number of poles. Most of the poles lie on the imaginary axis of the complex

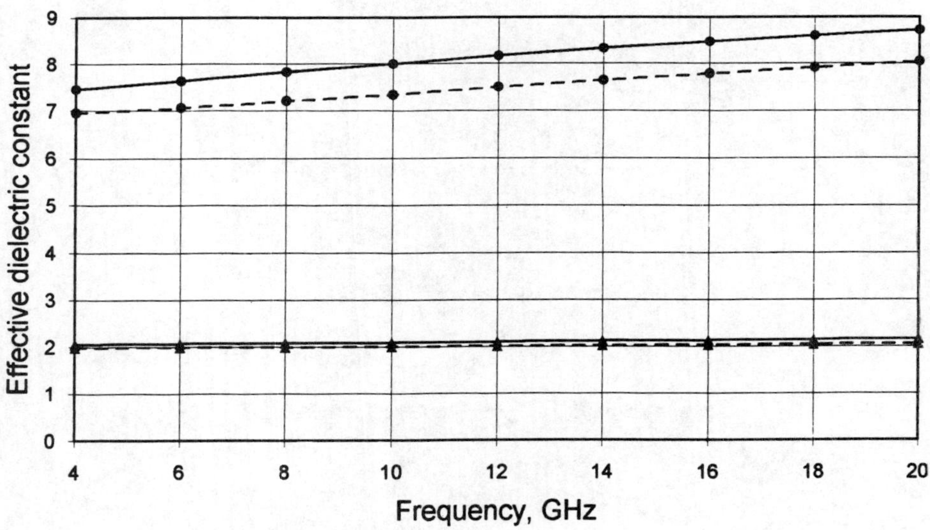

Figure 6.9 Effective dielectric constant versus frequency of a microstrip line: h = 0.1 cm; ----- w/h = 1.0,——w/h = 2.0, ●● ε_r = 10.0, ▲▲ ε_r = 2.5.

k_x-plane. If any pole exists on the real axis of the k_x-plane, one needs to use the singularity extraction technique, detailed in Appendix A2. However, this situation does not arise for the dimensions generally used in a microstrip line. Furthermore, the branch point also lie off the real axis of the k_x-plane because β is always greater than k_0. Therefore, the numerical integrations become straightforward, because no branch point exists on the integration path.

The propagation constant β is solved from (6.36) using a direct search algorithm. From β, the effective dielectric constant ε_{eff} is determined from the relation $\varepsilon_{\text{eff}} = \beta^2/k_0^2$. Figure 6.9 shows the variation of effective dielectric constant with frequency, with w/h and ε_r as parameters. The effective dielectric constant increases with the frequency of operation. This characteristic indicates the dispersive nature of a microstrip line. The dispersion effect is more prominent for substrates with larger dielectric constants.

6.4.4 The Characteristic Impedance

There are at least three definitions for the characteristic impedance of a microstrip line.[6] Here we use the following simple definition of the characteristic impedance, which is $Z_0 = V_L/I_L$, where I_L is the total current along the y direction, and V_L is the voltage between the strip and the ground plane. V_L can be expressed in terms of a line integral of the electric field between two conductors. For time-varying fields, the line integral is not unique and depends on the path chosen. To avoid the ambiguity, we take the weighted average value of the integral. The final result is

$$Z_0 = \frac{-\int_{-w/2}^{w/2} \int_{-h}^{0} E_z I_y \, dz \, dx}{\left[\int_{-w/2}^{w/2} I_y \, dx \right]^2} \tag{6.37}$$

Using (6.26b) for E_z and (6.24) for I_y, the expression for Z_0 is derived as

$$Z_0 = \frac{8\eta}{k_0 \varepsilon_r w^2 \pi} \int_{-\infty}^{\infty} \left[\beta |G(k_x)|^2 + \frac{A}{B} k_x F(k_x) G(k_x) \right] \bar{T}_m \, dk_x \tag{6.38}$$

[6]See for example, K. C. Gupta, R. Garg, and I. J. Bhal, *Microstrip Lines and Slot Lines*, Artech House, Dedham, MA, 1979, p. 49.

with

$$\bar{T}_m = \frac{y_{02}^m \tan(k_{1z}h)}{k_{1z}[y_{02}^m + jy_{01}^m \tan(k_{1z}h)]} \qquad \eta = 120\,\pi\,\Omega$$

The ratio A/B can be obtained from (6.34a). Figure 6.10 shows the variation of the characteristic impedance with frequency. The characteristic impedance increases with the frequency. As the frequency increases, the fringing effect decreases, so that the voltage beneath the strip conductor increases. This results in a larger characteristic impedance.

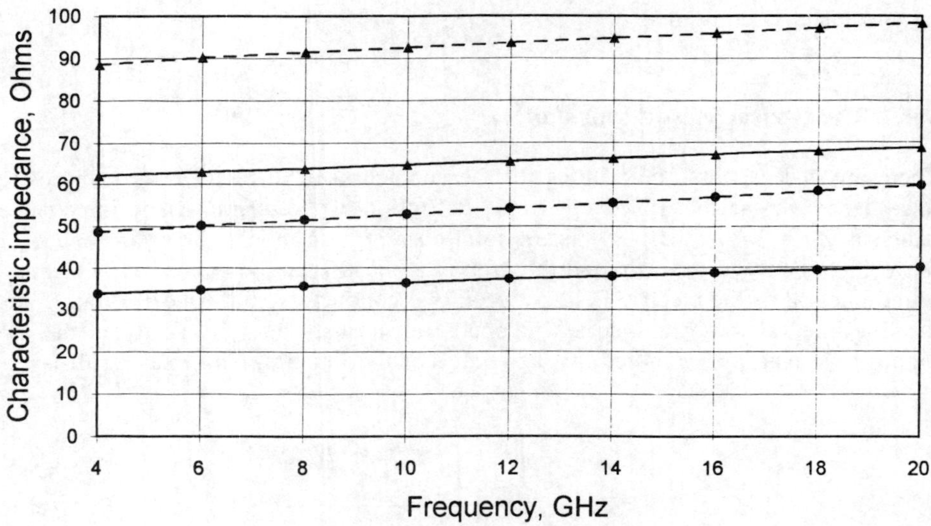

Figure 6.10 Characteristic impedance versus frequency of a microstrip line: h = 0.1 cm; ------ w/h = 1.0,——— w/h = 2.0, ●● ε_r = 10.0, ▲▲ ε_r = 2.5.

6.5 FULL WAVE ANALYSIS OF THE SLOT LINE

Slot line is another type of transmission line used in planar circuit. It consists of a narrow slot cut on a ground plane. Usually, one of the two sides of the ground plane is coated with a thin layer of dielectric substrate. A typical slot line structure is depicted in Figure 6.11.

In a slot line, the r.f. power propagates along the length of the slot. The dominant component of the electric field is transverse to the propagation direction. The

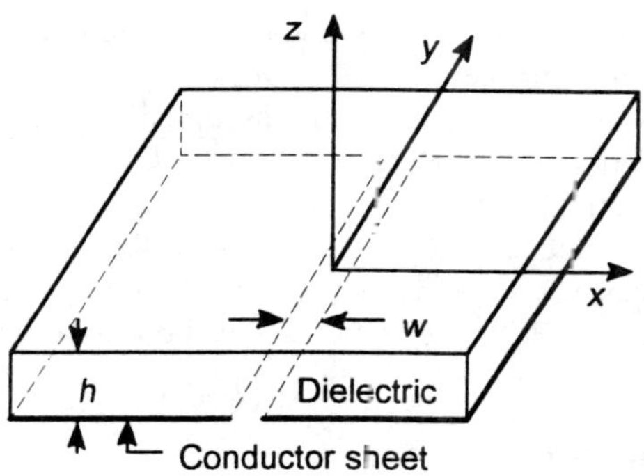

Figure 6.11 A slot line structure.

propagating mode in a slot line is hybrid in nature (neither purely TE_y nor TM_y). However, for low permittivity and electrically thin substrates, the propagation behavior is very close to that of a TEM mode.

The analysis of a slot line structure is very similar to that of a microstrip line, as detailed in Section 6.4. The boundary condition, however, is different from that of a microstrip line. To incorporate the effects of dielectric substrate and obtain the dispersion characteristics, one requires a full wave analysis of the structure. In the following subsections, the full wave analysis of a slot line is presented.

6.5.1 The Electric Field Distribution

To determine the r.f. characteristic of a slot line, we should find the electromagnetic fields components supported by the transmission line. As a first step toward that end, we assume the following trial field distributions on the slot plane:

$$E_x = \frac{A}{\sqrt{1 - \left(\frac{2x}{w}\right)^2}} \exp(-j\beta y) = Ag(x) \exp(-j\beta y) \tag{6.39a}$$

$$E_y = B\left(\frac{2x}{w}\right) \sqrt{1 - \left(\frac{2x}{w}\right)^2} \exp(-j\beta y) = Bf(x) \exp(-j\beta y) \tag{6.39b}$$

where A and B are two unknown constants; β is the propagation constant of the transmission line to be found. These field distributions are an electrostatic approximation of the field components on the slot opening that satisfy the edge conditions (see Problem 6.1). The slot width w is usually very small compared to the wavelength, therefore, the approximation yields fairly good accuracy. For more accurate result, we may use a number of terms in the assumed field distribution and then use the moment method formulation. Here, we will be limited to only one term, which is sufficient for all practical purposes.

From the aperture electric field, we can obtain the remaining field components on both sides of the aperture. The fields are contributed by the TM_z and TE_z modes. In the next subsection, the TM_z mode fields will be determined.

6.5.2 The TM_z Mode Field

To find the TM_z mode field components, we replace the aperture by an equivalent magnetic surface current. The equivalent magnetic surface current for the region $z > 0$ is given by

$$\mathbf{M} = \mathbf{E} \times \hat{z} = \hat{x}M_x + \hat{y}M_y$$

where

$$M_y = -Ag(x)\exp(-j\beta y) \tag{6.40a}$$

$$M_x = Bf(x)\exp(-j\beta y) \tag{6.40b}$$

The two-dimensional Fourier transforms of these magnetic surface current components are

$$\tilde{M}_x = BF(k_x)\delta(k_y - \beta) \tag{6.41a}$$

$$\tilde{M}_y = -AG(k_x)\delta(k_y - \beta) \tag{6.41b}$$

with

$$F(k_x) = \frac{j}{2}\frac{J_2(k_x w/2)}{k_x} \tag{6.42a}$$

$$G(k_x) = \frac{w}{4}J_0(k_x w/2) \tag{6.42b}$$

Using Table 2.1, we can construct the spectral domain equivalent circuit for the TM_z mode. The equivalent circuit is shown in Figure 6.12. The source current I_0 in the equivalent circuit is

$$I_0 = j(\hat{x}k_x + \hat{y}k_y) \times (\hat{x}\tilde{M}_x + \hat{y}\tilde{M}_y) \cdot \hat{z}$$

which after simplification becomes

$$I_0 = -j[Ak_xG(k_x) + Bk_xF(k_x)]\delta(k_y - \beta) \tag{6.43}$$

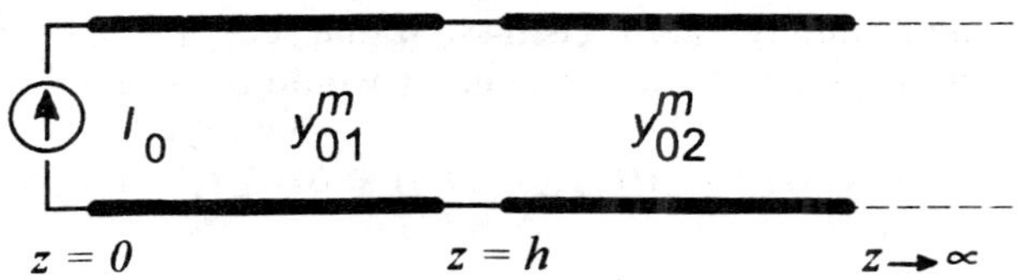

Figure 6.12 Spectral domain equivalent circuit of a slot line for the TM_z mode.

At $z = 0+$ the electric displacement vector component $\tilde{D}_z$ is given by $\tilde{D}_z(z = 0+) = I_0/y_{\text{in}}^m$, where

$$y_{\text{in}}^m = y_{01}^m \frac{y_{02}^m + jy_{01}^m \tan k_{1z}h}{y_{01}^m + jy_{02}^m \tan k_{1z}h}$$

For the present problem, the x and y components of the magnetic field are of importance. They can be expressed in terms of $\tilde{D}_z$ (see Section 2.2.4). At $z = 0+$, these components are derived as

$$H_x^m(z = 0+) = AH_{xA}^m(z = 0+) + BH_{xB}^m(z = 0+) \tag{6.44a}$$

$$H_y^m(z = 0+) = AH_{yA}^m(z = 0+) + BH_{yB}^m(z = 0+) \tag{6.44b}$$

with

$$H_{xA}^m(z = 0+) = -j\omega\beta \int_{-\infty}^{\infty} \frac{k_xG(k_x)}{(k_x^2 + \beta^2)y_{\text{in}}^m} \exp(-jk_xx)\, dk_x \exp(-j\beta y) \tag{6.45a}$$

$$H_{xB}^m(z = 0+) = -j\omega\beta^2 \int_{-\infty}^{\infty} \frac{F(k_x)}{(k_x^2 + \beta^2)y_{\text{in}}^m} \exp(-jk_xx)\, dk_x \exp(-j\beta y) \tag{6.45b}$$

$$H^m_{yA}\ (z = 0+) = j\omega \int_{-\infty}^{\infty} \frac{k_x^2 G(k_x)}{(k_x^2 + \beta^2) y^m_{\text{in}}} \exp(-jk_x x)\, dk_x \exp(-j\beta y) \tag{6.45c}$$

$$H^m_{yB}\ (z = 0+) = j\omega\beta \int_{-\infty}^{\infty} \frac{k_x F(k_x)}{(k_x^2 + \beta^2) y^m_{\text{in}}} \exp(-jk_x x)\, dk_x \exp(-j\beta y) \tag{6.45d}$$

In the preceding, the superscript m stands for the TM_z modes. The magnetic field components in the region $z < 0$ are obtained in a similar manner. The equivalent magnetic current at $z = 0-$ will be directed oppositely to that at $z = 0+$. Furthermore, because there is no dielectric layer, the spectral domain equivalent circuit consists of only one infinitely long transmission line section, with characteristic admittance of y^m_{02}. The magnetic field components would therefore be

$$H^m_x\ (z = 0-) = AH^m_{xA}\ (z = 0-) + BH^m_{xB}\ (z = 0-) \tag{6.46a}$$

$$H^m_y\ (z = 0-) = AH^m_{yA}\ (z = 0-) + BH^m_{yB}\ (z = 0-) \tag{6.46b}$$

where H^m_{xA} $(z = 0-)$ and so forth are obtainable from H^m_{xA} $(z = 0+)$ and so on, with y^m_{in} replaced by y^m_{02}. Also, an extra negative sign would appear with each expression to take care of the opposite polarity of the equivalent magnetic current at $z = 0-$.

6.5.3 The TE_z Mode Field

The TE_z mode fields are obtained from B_z. The spectral domain equivalent circuit for obtaining $\tilde{B}_z$ is shown in Figure 6.13. The voltage source V_0 in the equivalent circuit is obtained from Table 2.1, which after simplification becomes

$$V_0 = -\frac{1}{\omega} [\beta A G(k_x) - k_x B F(k_x)] \delta(k_y - \beta) \tag{6.47}$$

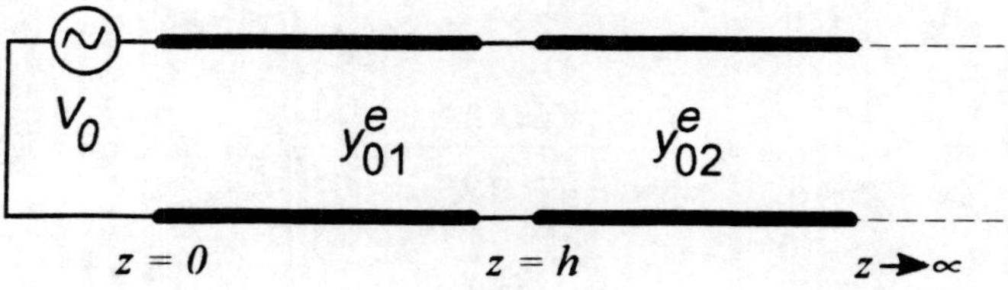

Figure 6.13 Spectral domain equivalent circuit of a slot line for the TE_z mode.

In the region $0 < z < h$, $\tilde{B}_z$ is given by

$$\tilde{B}_z = V_0 \frac{\exp(-jk_{1z}z) + \Gamma \exp(jk_{1z}z)}{(1 + \Gamma)}$$

with

$$\Gamma = \left[\frac{y_{01}^e - y_{02}^e}{y_{01}^e + y_{02}^e}\right] \exp(-2jk_{1z}h)$$

Using the results in Section 2.3.3, we can find H_x^e and H_y^e in terms of $\partial\tilde{B}_z/\partial z$. The final expression for the above quantities are

$$H_x^e\,(z = 0+) = AH_{xA}^e\,(z = 0+) + BH_{xB}^e\,(z = 0+) \tag{6.48a}$$

$$H_y^e\,(z = 0+) = AH_{yA}^e\,(z = 0+) + BH_{yB}^e\,(z = 0+) \tag{6.48b}$$

with,

$$H_{xA}^e\,(z = 0+) = \frac{-j\beta}{\omega} \int_{-\infty}^{\infty} \frac{k_x G(k_x)}{(\beta^2 + k_x^2)}\, y_{\text{in}}^e \exp(-jk_x x)\, dk_x \exp(-j\beta y) \tag{6.49a}$$

$$H_{xB}^e\,(z = 0+) = \frac{j}{\omega} \int_{-\infty}^{\infty} \frac{k_x^2 F(k_x)}{(\beta^2 + k_x^2)}\, y_{\text{in}}^e \exp(-jk_x x)\, dk_x \exp(-j\beta y) \tag{6.49b}$$

$$H_{yA}^e\,(z = 0+) = \frac{-j\beta^2}{\omega} \int_{-\infty}^{\infty} \frac{G(k_x)}{(\beta^2 + k_x^2)}\, y_{\text{in}}^e \exp(-jk_x x)\, dk_x \exp(-j\beta y) \tag{6.49c}$$

$$H_{yB}^e\,(z = 0+) = \frac{j\beta}{\omega} \int_{-\infty}^{\infty} \frac{k_x G(k_x)}{(\beta^2 + k_x^2)}\, y_{\text{in}}^e \exp(-jk_x x)\, dk_x \exp(-j\beta y) \tag{6.49d}$$

with

$$y_{\text{in}}^e = y_{01}^e \frac{y_{02}^e + jy_{01}^e \tan(k_{1z}h)}{y_{01}^e + jy_{02}^e \tan(k_{1z}h)}$$

The magnetic field components at $z = 0-$ can be obtained from the preceding expressions after replacing y_{in}^e by y_{02}^e and with an additional negative sign preceding each expression.

6.5.4 The Propagation Constant

The propagation constant β is determined from the magnetic field continuity on the slot opening. The total magnetic field must be continuous at $z = 0$, which yields

$$H_x^m(z = 0+) + H_x^e(z = 0+) = H_x^m(z = 0-) + H_x^e(z = 0-) \tag{6.50a}$$

$$H_y^m(z = 0+) + H_y^e(z = 0+) = H_y^m(z = 0-) + H_y^e(z = 0-) \tag{6.50b}$$

valid for $-w/2 < x < w/2$ and for all y. We now multiply the first equation by $f(x)$ and the second equation by $g(x)$ and then integrate with respect to x from $-w/2$ to $w/2$. The final result is

$$AX_1 + BY_1 = 0 \tag{6.51a}$$

$$AX_2 + BY_2 = 0 \tag{6.51b}$$

with

$$\begin{aligned} X_1 &= t_1 + t_2 \qquad & Y_1 &= t_3 + t_4 \\ X_2 &= t_3 + t_4 \qquad & Y_2 &= t_5 + t_6 \end{aligned} \tag{6.52}$$

where

$$t_1 = 2\pi j\omega \int_{-\infty}^{\infty} \frac{k_x^2}{k_x^2 + \beta^2} |G(k_x)|^2 \left[\frac{1}{y_{\text{in}}^m} + \frac{1}{y_{02}^m}\right] dk_x \tag{6.53a}$$

$$t_2 = \frac{-2\pi j\beta^2}{\omega} \int_{-\infty}^{\infty} \frac{|G(k_x)|^2}{k_x^2 + \beta^2} [y_{\text{in}}^e + y_{02}^e]\, dk_x \tag{6.53b}$$

$$t_3 = 2\pi j\omega\beta \int_{-\infty}^{\infty} \frac{k_x F(k_x) G(k_x)}{k_x^2 + \beta^2} \left[\frac{1}{y_{\text{in}}^m} + \frac{1}{y_{02}^m}\right] dk_x \tag{6.53c}$$

$$t_4 = \frac{2\pi j\beta}{\omega} \int_{-\infty}^{\infty} \frac{k_x F(k_x) G(k_x)}{k_x^2 + \beta^2} [y_{\text{in}}^e + y_{02}^e]\, dk_x \tag{6.53d}$$

$$t_5 = -2\pi j\omega\beta^2 \int_{-\infty}^{\infty} \frac{|F(k_x)|^2}{k_x^2 + \beta^2} \left[\frac{1}{y_{\text{in}}^m} + \frac{1}{y_{02}^m}\right] dk_x \tag{6.53e}$$

$$t_6 = \frac{2\pi j}{\omega} \int_{-\infty}^{\infty} \frac{k_x^2 \, |F(k_x)|^2}{k_x^2 + \beta^2} \, [y_{\text{in}}^e + y_{02}^e] \, dk_x \tag{6.53f}$$

For nontrivial solutions for A and B, the equations must be consistent implying

$$X_1Y_2 - X_2Y_1 = 0 \tag{6.54}$$

This equation is used to find the unknown propagation constant β. A numerical search technique is invoked to find β. The integrals are computed using a numerical integration routine. Each of the previous infinite integrals was truncated to a very large value of k_x. For the remaining range of k_x values the integrals are obtained asymptotically. For small substrate thickness, no singularity exists on the real axis of k_x, which makes the numerical integration straightforward. The effective dielectric constant, ε_{eff}, is obtained from relation $\varepsilon_{\text{eff}} = \beta^2/k_0^2$. Figure 6.14 shows the variation of ε_{eff} with the substrate thickness h. As expected, ε_{eff} increases with h. For a given h, ε_{eff} is larger for a smaller w.

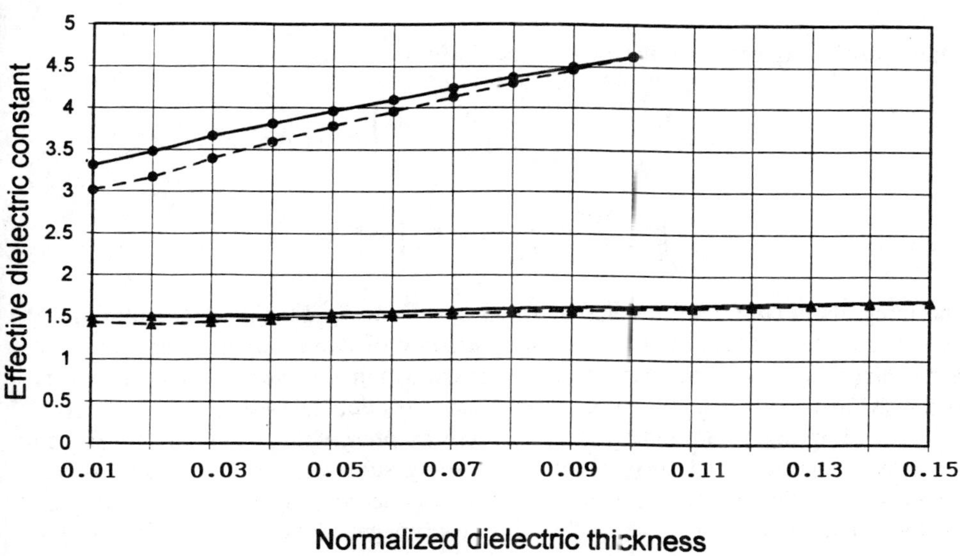

Figure 6.14 Effective dielectric constant of a slot line versus normalized dielectric thickness h/λ_0:——$w/h = 0.25$, ----- $w/h = 0.5$, ● $\varepsilon_r = 8.0$, ▲ $\varepsilon_r = 2.5$.

6.5.5 The Characteristic Impedance

The characteristic impedance may be obtained from the total power that flows across the cross section of the slot line. This necessitates integration of the Poynting vector on the cross-sectional area. However, for small dielectric thickness, the propagating mode is very close to the TEM_y mode, and the characteristic impedance can be approximated as

$$Z_0 = V/I_y \tag{6.55}$$

where V is the voltage at $y = 0$ across the slot, which is

$$V = -\int_{-w/2}^{w/2} E_x\, dx = -A\int_{-w/2}^{w/2} g(x)\, dx = \frac{-\pi}{2} wA \tag{6.56}$$

The current that flows in one conductor surface is given by

$$I_y = \int_{w/2}^{\infty} H_x(z = 0+)\, dx - \int_{w/2}^{\infty} H_x(z = 0-)\, dx$$

After substituting expressions for H_x, one derives

$$\begin{aligned} I_y = -\Bigg[2\omega\beta \int_0^{\infty} \frac{G(k_x)}{(k_x^2 + \beta^2)} \left[\frac{1}{y_{\text{in}}^m} + \frac{1}{y_{02}^m}\right] \cos(k_x w/2)\, dk_x \\ + \frac{2\beta}{\omega} \int_0^{\infty} \frac{G(k_x)}{(k_x^2 + \beta^2)} \left[y_{\text{in}}^e + y_{02}^e \right] \cos(k_x w/2)\, dk_x \Bigg] A \end{aligned} \tag{6.57}$$

The integrals in (6.57) were computed numerically to find the characteristic impedance of slot lines. Figure 6.15 shows the variation of Z_0 versus the normalized substrate thickness, h/λ_0. The characteristic impedance initially increases with h, reaches to a maximum and then decreases. The value of Z_0 decreases with w/h.

It is important to mention that the analysis presented in this chapter is valid for a lossy dielectric substrate, too. For a lossy substrate, the dielectric constant should be considered a complex number, with a nonzero imaginary part representing the loss. The effect of conductor loss on the propagation constant can be included to some extent using a perturbational method.[7]

---

[7]R. F. Harrington, *Time-Harmonic Electromagnetic Fields*, McGraw-Hill, New York, 1961, pp. 72–73.

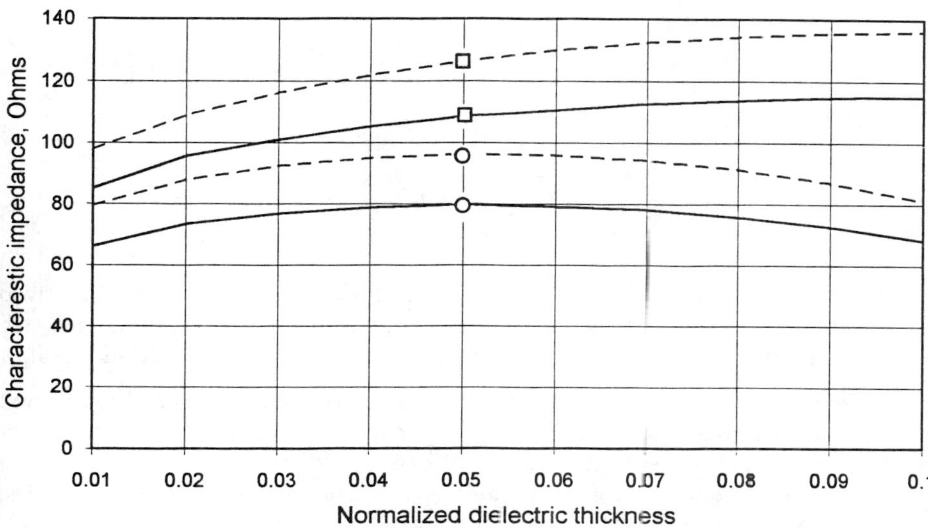

Figure 6.15 Characteristic impedance of a slot line versus normalized dielectric thickness h/λ_0:—— $w/h = 0.25$, ----- $w/h = 0.5$, —○— $\varepsilon_r = 8.0$, —□— $\varepsilon_r = 2.5$.

PROBLEMS

6.1 Verify that in the cylindrical coordinate system (ρ, θ, z), $\Phi = \sin(\theta/2)/\sqrt{\rho}$ is a solution of the Laplace equation $\nabla^2\Phi = 0$. Note that the zero potential surface in this case is a half-plane extended from $\rho = 0$ to ∞ at $\theta = 0$. The surface has a sharp edge at $\rho = 0$. Show that the surface charge distribution on the equipotential (zero potential) surface is proportional to $1/\sqrt{\rho}$, justifying the assumed charge distribution in (6.1). (Hints: determine E_θ at $\theta = 0$ and $\theta = 2\pi$. From discontinuity of E_θ, calculate the surface charge distribution.)

6.2 Show that the Fourier transformation of the function $T_n(2x/w)/\sqrt{1 - (2x/w)^2}$ is given by

$$\frac{1}{2\pi}\int_{-w/2}^{w/2} \frac{T_n(2x/w)}{\sqrt{1 - (2x/w)^2}} \exp(jk_x x)\, dx = \frac{w}{4} j^n J_n(k_x w/2)$$

6.3 Obtain the capacitance per unit of length of a microstrip line using spectral Galerkin's method. Use (6.9) for the potential function. Because $\Phi(x, 0)$ should be constant for $-w/2 < x < w/2$, and for $n \neq 0$,

$$\int_{-w/2}^{w/2} T_n(2x/w)/\sqrt{1-(2x/w)^2}\, dx = 0$$

one should have

$$\int_{-w/2}^{w/2} \frac{\Phi(x, 0)\, T_{2i}(2x/w)}{\sqrt{1-(2x/w)^2}}\, dx = 0 \qquad i \neq 0$$

Show that for $i = 1, 2, \ldots, N - 1$, this equation set is same as that given in (6.13). Hence, for this particular choice of the expansion functions, the variational formulation and the spectral Galerkin's technique are identical.

6.4 Assuming a steady current on the strip conductor, find the general expression for the inductance per unit of length for a microstrip line. For the current distribution, use the same expansion function as given in (6.1). (Hints: For the steady electric current, the mode is TE_z type. Solve for B_z using the TE_z equivalent circuit in the spectral domain.)

6.5 Derive (6.17) through (6.19).

6.6 Derive (6.20) through (6.22).

6.7 Show that the integral

$$I(s) = \int_0^\infty J_n^2\,(k_x w/2) \cos^2\left(k_x \frac{s+w}{2} - \frac{n\pi}{2}\right) \frac{\tanh k_x h}{k_x(\varepsilon_r + \tanh k_x h)}\, dk_x$$

approaches

$$\frac{1}{2}\int_0^\infty J_n^2\,(k_x w/2) \frac{\tanh k_x h}{k_x(\varepsilon_r + \tanh k_x h)}\, dk_x$$

as $s \to \infty$. Hence, prove that, for the even mode, the characteristic impedance of a coupled line approaches to the characteristic impedance of an isolated microstrip line as the separation between the coupled lines increases. The result is also true for odd-mode impedance.

6.8 Figure 6.16 shows a simple coupled line circuit. The even- and odd-mode characteristic impedances of the coupled line are 53 Ω and 42.5 Ω, respectively. The corresponding effective dielectric constants are 2.20 and 1.95, respectively. The length of the coupled line is 3 cm. Find the power in the terminated loads, if the source voltages are $V_1 = 1$ V and $V_2 = 2$ V. The frequency of operation is 2 GHz.

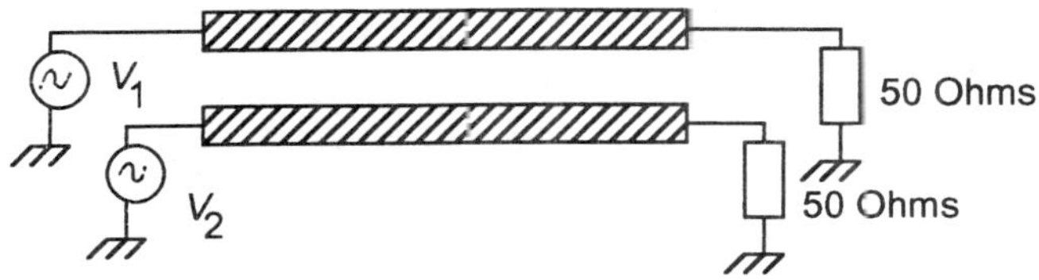

Figure 6.16 A microwave circuit of coupled lines.

6.9 Verify (6.35a) through (6.35f).

6.10 Formulate the propagation constant and characteristic impedance of a shielded microstrip line as shown in Figure 6.17. The strip width is w. The dielectric thickness is h with dielectric constant of ε_r. The waveguide dimensions are $a \times b$.

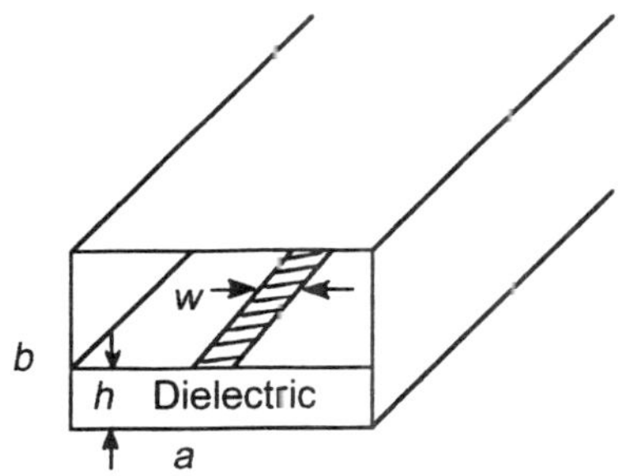

Figure 6.17 A shielded microstrip line.

6.11 Perform a full wave analysis of a coupled line shown in Figure 6.5. Assume a current distribution on each strip as given in (6.23) and (6.24). Find characteristic impedances and propagation constants for both even and odd modes.

6.12 Show that for a slot line the propagation constant β is equal to k_0 if the dielectric constant of the substrate ε_r is equal to unity. Also show that, in such a case, the propagating mode is TEM_y mode with $H_y = 0$.

6.13 Determine the static capacitance per unit of length for a slot line shown in Figure 6.11. Assume a uniform trial electric field across the slot. (Hints: From the equivalent magnetic current, find D_z. The surface charge density on a conductor is $\rho_s = D_z$. Use $C = Q/V$ formula.)

6.14 Obtain the expression for effective dielectric constant and characteristic impedance of a slot line using static formulation.

6.15 Figure 6.18 shows the structure of a fin line. Assume the field distribution across the slot is same as given in (6.39). Find the propagation constant and the characteristic impedance.

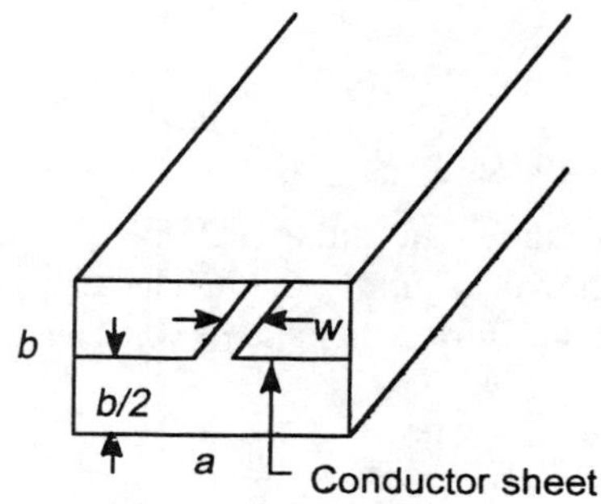

Figure 6.18 A fin line structure.

Appendix 1
The Saddle Point Method of Integration

The saddle point method is used to obtain the asymptotic expression of a function when the function has the following integral form:

$$F(\lambda) = \int_c g(z) \exp\{\lambda f(z)\}\, dz \tag{A1.1}$$

where λ is real, z is complex, and c is the integration path on the complex z-plane (see Figure A1.1). We would like to obtain the asymptotic expression for $F(\lambda)$ when λ is very large. When λ is large and positive, the value of the exponential becomes large if the real part of $f(z)$ is positive. As λ increases indefinitely, the entire contribution of the exponential to the integrand will come from the region in which the real part of $f(z)$ takes a positive maximum value. Away from this region, the contribution will be insignificant in comparison. Let us express $f(z)$ as $f(z) = u(x, y) + jv(x, y)$, where $z = x + jy$, x and y are real variables, u and v are real functions. Then $F(\lambda)$ can be expressed as

$$F(\lambda) = \int_c g(z) \exp\{\lambda u(x, y)\} \exp\{j\lambda v(x, y)\}\, dz \tag{A1.2}$$

Clearly, for λ large and positive and for $u(x, y) > 0$, the dominant factor in the integrand is the factor $\exp\{\lambda u(x, y)\}$. The other exponential factor has a purely imaginary argument, so that the magnitude of that factor is unity.

Now suppose, in the vicinity of the path of integration there exists a point P at $z = z_0$ such that $f'(z_0) = 0$, where the prime denotes the derivative with respect to z. For the analytic function $f(z)$, we employ the Taylor series expansion as follows:

$$f(z) = f(z_0) + \frac{(z - z_0)^2}{2!} f''(z_0) + \frac{(z - z_0)^3}{3!} f'''(z_0) + \ldots \tag{A1.3}$$

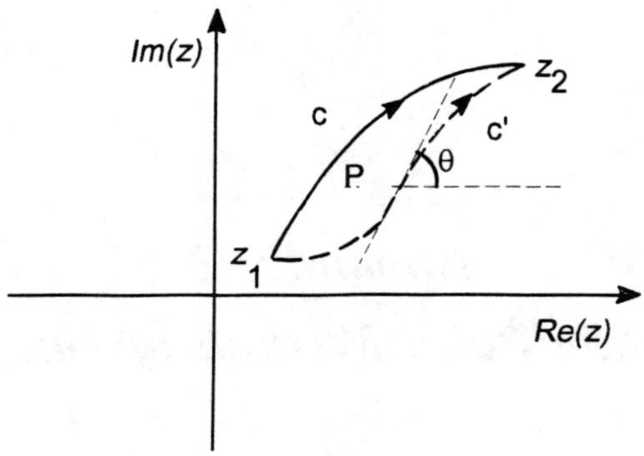

Figure A1.1 Path for saddle point integration: ________ original path, ------ deformed path.

The point P is known as the *saddle point* for the following reason. Because $f'(z) = 0$ at P, both functions $u(x, y)$ and $v(x, y)$ have extreme values at that point. Moreover, in the vicinity of the point P, each surface looks like a saddle, having a maximum in one direction, and a minimum in its orthogonal direction. This can be proven from Cauchy-Riemann conditions,[1] which is given by $\partial u/\partial x = \partial v/\partial y$, $\partial u/\partial y = -\partial v/\partial x$. From these conditions, it is straight forward to prove that

$$\frac{\partial^2 u}{\partial x^2} + \frac{\partial^2 u}{\partial y^2} = 0 = \frac{\partial^2 v}{\partial x^2} + \frac{\partial^2 v}{\partial y^2} \tag{A1.4}$$

Now if u passes through a maximum along the x direction, then $\partial^2 u/\partial x^2 < 0$. From (A1.4), we would have $\partial^2 u/\partial y^2 > 0$, implying that u must pass through a minimum along the y direction, which is orthogonal to x direction. This is in general true for any two mutually orthogonal directions, because the form of (A1.4) does not change in any other orthogonal coordinate system. The surface $v = v(x, y)$ also follows a similar characteristic.

Returning to our original problem, we substitute $f(z)$ in terms of the Taylor series expansion and retain the first two terms. This yields

$$F(\lambda) \approx \int_c g(z) \exp\left\{ \lambda f(z_0) + \frac{\lambda(z - z_0)^2}{2} f''(z_0) \right\} dz \tag{A1.5}$$

We now substitute $(z - z_0) = \rho \exp(j\theta)$, yielding

[1]See E. Kreyszig, Advanced Engineering Mathematics, 6th ed., John Wiley, New York, 1988, p. 740.

$$F(\lambda) \approx \int_c g\{z_0 + \rho \exp(j\theta)\} \exp\{\lambda f(z_0)\} \exp\left\{\frac{\lambda}{2} f''(z_0)\rho^2 \exp(2j\theta)\right\} dz$$

We write $f''(z_0) = |f''(z_0)| \exp(j\phi)$. Also, assume that $g(z)$ is a slowly varying function of z. Therefore, the above equation becomes

$$F(\lambda) \approx g(z_0) \exp\{\lambda f(z_0)\} \int_c \exp\left[\frac{\lambda}{2} \rho^2 |f''(z_0)| \exp(2j\theta + j\phi)\right] dz \qquad \text{(A1.6)}$$

After separating the real and imaginary parts of the exponential term inside the square bracket, we obtain

$$F(\lambda) \approx g(z_0) \exp\{\lambda f(z_0)\} \int_c \exp\left\{\rho^2 \frac{\lambda}{2} |f''(z_0)| \cos(2\theta + \phi)\right\} \times \exp\left\{j\rho^2 \frac{\lambda}{2} |f''(z_0)| \sin(2\theta + \phi)\right\} dz \qquad \text{(A1.7)}$$

We now perform the integration along an alternate path c'. The path c' passes through P and has endpoints z_1 and z_2. In the vicinity of P, the value of θ remains constant along c' (see Figure A1.1). If $f(z)$ and $g(z)$ are analytic in the closed region formed by c and c', the value of the integral along c' will be identical with that along c. Also, we choose θ such that $\sin(2\theta + \phi) = 0$, which yields $\theta = (n\pi - \phi)/2$, $n = 0, \pm 1, \pm 2, \ldots$

One of these values of θ makes the second exponential factor equal to unity along the segment of c' near P. To make the first exponential factor passing through a maximum at $z = z_0$ (that is, when $\rho = 0$), we should select θ such that $\cos(2\theta + \phi) = -1$, or

$$\theta = \frac{\pm\pi - \phi}{2} \qquad \text{(A1.8)}$$

In the vicinity of P, θ is constant along c. Hence, we write $dz = \exp(j\theta)\, d\rho$.

Therefore, (A1.7) reduces to

$$F(\lambda) \approx g(z_0) \exp\{\lambda f(z_0)\} \exp(j\theta) \int_\rho \exp\left\{-\rho^2 \frac{\lambda}{2} |f''(z_0)|\right\} d\rho$$

For very large λ, the integration can be performed over the limit of ρ from $-\infty$ to ∞, because the integrand is negligibly small away from $\rho = 0$ point. Therefore, we write

$$F(\lambda) \approx g(z_0) \exp\{\lambda f(z_0) + j\theta\} \int_{-\infty}^{\infty} \exp\left\{-\rho^2 \frac{\lambda}{2} |f''(z_0)|\right\} d\rho$$

To evaluate the integral, we now use the following well-known result[2] $\int_{-\infty}^{\infty} \exp(-x^2/2\sigma^2)\, dx = \sqrt{2\pi}\sigma$. This yields

$$F(\lambda) \approx g(z_0) \exp\{\lambda f(z_0) + j\theta\} \sqrt{\frac{2\pi}{\lambda |f''(z_0)|}} \tag{A1.9}$$

where θ is given in (A1.8). It is important to note that (A1.8) yields two different values of θ, 180° apart. The correct value of θ is the one that matches with the angle between the real axis and the path c' at P (see the following example). It is to be pointed out that, if any singularity exists in the region enclosed by c and c', the contribution of the singular point must be included together with the contribution of the saddle point.

In some cases, more then one saddle point may exist in the vicinity of an integration path. In such cases, consideration from all saddle points must be added to obtain the final asymptotic value.

Example. The integral form of the nth order Bessel function is given by

$$J_n(\lambda) = \frac{j^{-n}}{2\pi} \int_{-\pi+\varepsilon}^{\pi+\varepsilon} \exp(jnz) \exp(j\lambda \cos z)\, dz$$

Find an asymptotic expression for $J_n(\lambda)$ when λ is large and positive.

Solution. The original path c is along the real axis of the z-plane as shown in Figure A1.2. Now, comparing with (A1.1), we have for the present problem

$$g(z) = \frac{j^{-n}}{2\pi} \exp(jnz) \qquad f(z) = j \cos z$$

Now $f'(z_0) = 0$ implies $z_0 = 0, \pi$ are two saddle points exist in the vicinity of the original contour c (in fact the saddle points lie on c). We have to find the contributions from each saddle point.

Consider the saddle point $z_0 = 0$. At that point we obtain

$$g(z_0) = \frac{j^{-n}}{2\pi} \qquad f(z_0) = j$$

[2]See ibid., p. 1211.

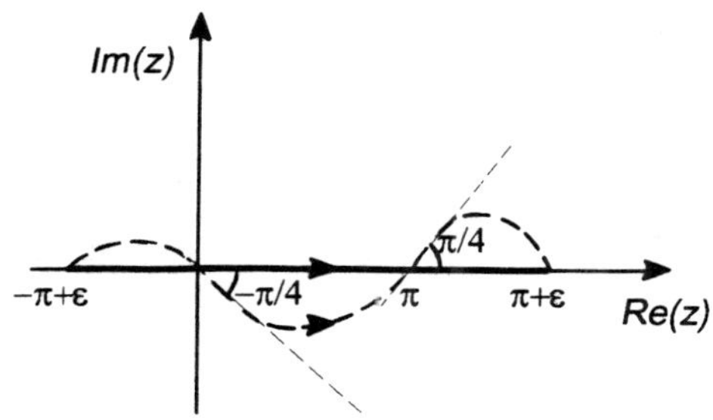

Figure A1.2 Deformed path for saddle point integration. The original path is along the real axis. The deformed path is the dashed curve.

$$f''(z_0) = -j \qquad |f''(z_0)| = 1 \qquad \phi = -\pi/2$$

$$\theta = \frac{\pm\pi - \phi}{2} = \frac{-\pi}{4}, \frac{3\pi}{4}$$

The appropriate value for θ should be $-\pi/4$. The other value for θ is not possible (see Figure A1.2).

Therefore, for the saddle point at $z = 0$, the contribution is obtained using (A1.9), which is

$$I_1 = \frac{j^{-n}}{2\pi} \exp\left\{j\left(\lambda - \frac{\pi}{4}\right)\right\} \sqrt{\frac{2\pi}{\lambda}}$$

Now consider the other saddle point at $z = \pi$. We obtain

$$g(z_0) = \frac{j^{-n}}{2\pi} \exp(jn\pi)$$

$$f(z_0) = -j$$

$$f''(z_0) = j \qquad |f''(z_0)| = 1 \qquad \phi = \pi/2$$

$$\phi = \frac{\pm\pi - \phi}{2} = \frac{\pi}{4}, -\frac{3\pi}{4}$$

Referring Figure A1.2, the correct choice of θ at $z = \pi$ would be $\pi/4$. The contribution of the saddle point at $z = \pi$ therefore is

$$I_2 = \frac{j^{-n}}{2\pi} \exp(jn\pi) \exp\left\{j\left(-\lambda + \frac{\pi}{4}\right)\right\} \sqrt{\frac{2\pi}{\lambda}}$$

For large λ the asymptotic expression for $J_n(\lambda)$ is

$$J_n(\lambda) \approx I_1 + I_2 = \frac{j^{-n}}{2\pi} \sqrt{\frac{2\pi}{\lambda}} \left\{\exp(jn\pi) \exp\left(-\lambda j + j\frac{\pi}{4}\right) + \exp\left(\lambda j - j\frac{\pi}{4}\right)\right\}$$

After simplification, the final expression becomes

$$J_n(\lambda) \approx \sqrt{\frac{2}{\pi\lambda}} \cos\left(\lambda - \frac{n\pi}{2} - \frac{\pi}{4}\right)$$

Practice Problem. Show that the asymptotic expression for $F(\lambda)$ is $\sqrt{2\pi\lambda}\, e^{-\lambda}\lambda^{\lambda}$ if $F(\lambda) = \int_0^\infty e^{-x}x^{\lambda}dx$.

Appendix 2
The Singularity Extraction Technique

The singularity extraction technique facilitates numerical computation of a complex line integral if the integrand has a pole located at an infinitesimal distance from the integration path (not exactly on the path). An integral with a singular point may be expressed as

$$I = \int_{z_1}^{z_2} \frac{f(z)}{(z - z_0)} dz \tag{A2.1}$$

where $f(z)$ is analytic everywhere on the path of integration; z_1 and z_2 denote two endpoints of the integration path and are complex numbers; z_0 is the point of singularity that lies in the close vicinity of the integration path. Because z_0 is located at an infinitesimal distance from the path of integration, numerical solution of the integral becomes unstable. To circumvent the situation, we express the integral in the following way:

$$I = \int_{z_1}^{z_2} \frac{f(z) - f(z_0)}{(z - z_0)} dz + f(z_0) \int_{z_1}^{z_2} \frac{1}{(z - z_0)} dz \tag{A2.2}$$

The first integral in (A2.2) has no singularity at $z = z_0$. Therefore, any numerical integration scheme (such as Simpson's rule) can be invoked to evaluate this integral. To evaluate the integrand at $z = z_0$, one may use L'Hospital's rule, which turns out to be $f'(z_0)$, in this case. The second integral, however, is interesting. Because, z_0 does not lie "exactly" on the path, the integration can be performed analytically. In fact, the second integral would have a closed form expression, which is

$$I_2 = f(z_0) \int_{z_1}^{z_2} \frac{1}{(z - z_0)} dz = f(z_0)\{\ln(z_2 - z_0) - \ln(z_1 - z_0)\}$$

The final expression for I becomes

$$I = \int_{z_1}^{z_2} \frac{f(z) - f(z_0)}{(z - z_0)} dz + f(z_0)\{\ln(z_2 - z_0) - \ln(z_1 - z_0)\} \tag{A2.3}$$

It is important to note that the logarithmic functions are multivalued functions in the complex domain. The numerical value of I_2 depends on the relative location of the pole with respect to the path of integration. This point is illustrated in the following example.

Example. Consider a magnetic line current over a dielectric-coated ground plane. The displacement vector component D_z in the air region is given by (see (3.5) in Chapter 3)

$$D_z = \frac{-jM_0}{2\pi} \int_{-\infty}^{\infty} \frac{k_y}{y_{\text{in}}^m} \exp(-jk_{0z}z) \exp(-jk_y y)\, dk_y \tag{A2.4}$$

with

$$y_{\text{in}}^m = \frac{jk_{0z}}{\varepsilon_0} - \frac{k_{1z} \tan k_{1z}h}{\varepsilon_0 \varepsilon_r} \qquad k_{0z} = \sqrt{k_0^2 - k_y^2} \qquad k_{1z} = \sqrt{k_1^2 - k_y^2}$$

For electrically thin dielectric, y_{in}^m has two zeros near the real k_y axis at $\pm k_{y0}$. Use the singularity extraction technique to find an alternative expression for D_z that is suitable for numerical computations.

Solution. Equation (A2.4) can be expressed as

$$D_z = \frac{-M_0}{\pi} \int_0^{\infty} \frac{k_y \sin(k_y y)}{y_{\text{in}}^m} \exp(-jk_{0z}z)\, dk_y \tag{A2.5}$$

Now on the integration path $0 < k_y < \infty$, there is only one zero of y_{in}^m. Also, the zero lies between k_0 and k_1. We can write D_z as

$$\begin{aligned} D_z &= \frac{-M_0}{\pi} \left[\int_0^{k_0} + \int_{k_0}^{k_1} + \int_{k_1}^{\infty} \right] \frac{k_y \sin(k_y y)}{y_{\text{in}}^m} \exp(-jk_{0z}z)\, dk_y \\ &= \frac{-M_0}{\pi} (I_1 + I_2 + I_3) \end{aligned} \tag{A2.6}$$

The first integral and the last integral have no singular point near the path of integration. Therefore, they can be computed numerically without any problem. The second integral, however, has a singular point near the path. Let us focus on this integral.

$$I_2 = \int_{k_0}^{k_1} \frac{k_y \sin(k_y y)}{y_{\text{in}}^m} \exp(-jk_{0z} z)\, dk_y \tag{A2.7}$$

Let us define $f(k_y)$ as

$$f(k_y) = \frac{k_y \sin(k_y y)}{y_{\text{in}}^m} (k_y - k_{y0}) \exp(-jk_{0z} z) \tag{A2.8}$$

Then, I_2 can be expressed as

$$I_2 = \int_{k_0}^{k_1} \frac{f(k_y)}{(k_y - k_{y0})}\, dk_y \tag{A2.9}$$

Now, $f(k_y)$ is analytic everywhere in the interval $k_0 < k_y < k_1$. From (A2.3), we can write

$$I_2 = \int_{k_0}^{k_1} \frac{f(k_y) - f(k_{y0})}{(k_y - k_{y0})}\, dk_y + f(k_{y0}) \{\ln(k_1 - k_{y0}) - \ln(k_0 - k_{y0})\} \tag{A2.10}$$

It is important to note that the last logarithmic term has a negative argument, because $k_{y0} > k_0$. This term can be expressed as

$$f(k_{y0}) \ln(k_0 - k_{y0}) = f(k_{y0}) \ln(k_{y0} - k_0) + f(k_{y0}) \ln(-1) \tag{A2.11}$$

Now $\ln(-1)$ is multivalued. The general expression is $\ln(-1) = \pm j\pi$, $\pm j3\pi$, $\pm j5\pi$, and so on. The choice depends on the physical problem. For the present case, k_{y0} is considered to have an infinitesimal negative imaginary part. Therefore, when k_y goes from the left to the right of k_{y0}, the angle of $(k_y - k_{y0})$ changes from π to 0 (see Figure A2.1). Therefore,

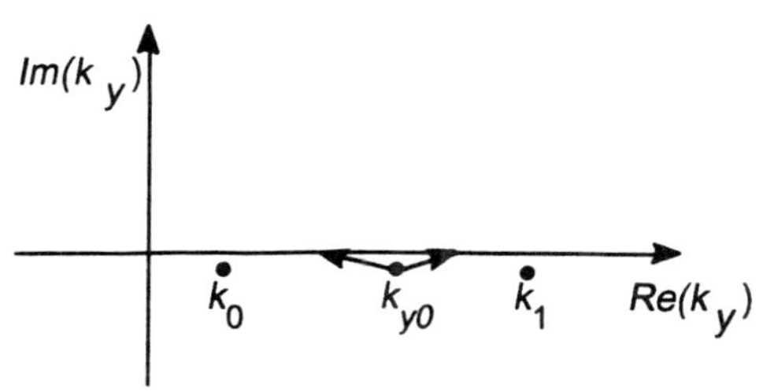

Figure A2.1 The rotation of the vector $(k_y - k_{y0})$ when k_y moves from the left to the right of $k_{y0} \cdot k_{y0}$ has an infinitesimal negative imaginary part, k_0 and k_1 also have infinitesimal negative imaginary parts.

$$\int_{k_{y0-}}^{k_{y0+}} \frac{1}{k_y - k_{y0}}\, dk_y = -j\pi \tag{A2.12}$$

Considering this result in (A2.9), we would have

$$I_2 = \int_{k_0}^{k_1} \frac{f(k_y) - f(k_{y0})}{(k_y - k_{y0})}\, dk_y + f(k_{y0}) \ln\frac{(k_1 - k_{y0})}{(k_{y0} - k_0)} - j\pi f(k_{y0}) \tag{A2.13}$$

where $f(k_{y0})$ can be determined from (A2.8), using L'Hospital's rule. The final expression for D_z therefore becomes

$$\begin{aligned} D_z = {} & \frac{-M_0}{\pi}\left[\int_0^{k_0} + \int_{k_1}^{\infty}\right] \frac{f(k_y)}{k_y - k_{y0}}\, dk_y - \frac{M_0}{\pi}\int_{k_0}^{k_1} \frac{f(k_y) - f(k_{y0})}{(k_y - k_{y0})}\, dk_y \\ & - \frac{M_0}{\pi} f(k_{y0}) \left[\ln\frac{(k_1 - k_{y0})}{(k_{y0} - k_0)} - j\pi\right] \end{aligned} \tag{A2.14}$$

This expression for D_z is suitable for numerical computation because no singular point exists on the path of integrations.

Appendix 3
Multiple-Valued Functions, Branch Point, and Branch Cut

Multiple-Valued Functions and Branches

A function $f(z)$ is said to be multiple valued if more then one value of $f(z)$ corresponds to each value of z. Consider the function $f(z) = \sqrt{z}$. As shown in Figure A3.1, suppose z makes a complete counterclockwise rotation around the origin of the complex z-plane starting from the point A. At the point A, z can be expressed as $z_A = r\,\exp(j\theta_1)$, so that $f(z_A) = \sqrt{r}\,\exp(j\theta_1/2)$. After a 360° rotation, the point z comes back to the same point A. After a 360° rotation of z along the contour, the angle of z becomes $\theta_1 + 2\pi$, so that $f(z_A)$ becomes $\sqrt{r}\,\exp(j\theta_1/2 + j\pi) = -\sqrt{r}\,\exp(j\theta_1/2)$. Therefore, two distinct values can be assigned for $f(z)$ at the same point A. Hence, $f(z)$ is a multivalued function, in this case a two-valued function.

Now, if $0 < \theta < 2\pi$, $f(z)$ is single valued for all θ. Let us call $f(z)$ as $f_1(z) = r\,\exp(j\theta/2)$ for $0 < \theta < 2\pi$. Also call $f(z) = f_2(z) = r\,\exp(j\theta/2 + j\pi)$ for $2\pi <$

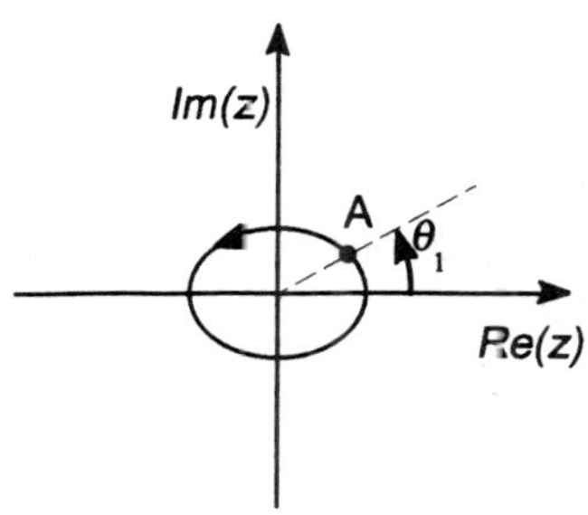

Figure A3.1 A closed contour around the origin of the complex z-plane.

$\theta < 4\pi$. Therefore, $f_1(z)$ and $f_2(z)$ are single-valued functions in their defined domain, $f_1(z)$ and $f_2(z)$ are called *branches* of the two-valued function $f(z)$.

Branch Point and Branch Cut

A branch cut is the boundary of the domains of the branches of a multivalued function. In the example $f(z) = \sqrt{z}$, the boundary of the two branches is the line $\theta = 2\pi$ (the positive real axis).[1] The branch point is the point in the complex z-plane, around which point a function $f(z)$ exhibits a multivalued nature. For this example, the origin ($z = 0$) is the branch point.

Integration Around a Branch Cut

In many situations, one may have to compute an integral around a branch cut. From computational point of view, it is important to simplify the integral, which will reduce the computation effort considerably. Consider the following integral

$$I = \int_c \frac{f(z)}{b + \sqrt{z - a}} dz$$

where the path of integration around the branch cut is shown in Figure A3.2. Here, a and b are complex numbers; $f(z)$ is an analytic function. The branch point exists at $z = a$. The branch cut starts from the branch point at $z = a$ and extends to a point on the circle of infinite radius. We can decompose the integral as

$$I = \left[\int_B^A + \int_A^{A'} + \int_{A'}^{B'} \right] \frac{f(z)}{\sqrt{z - a} + b} dz$$

$$= I_1 + I_2 + I_3$$

which essentially is a summation of three integrals. We will show that, upon simplification, these three integrals reduce to a single integral.

First consider the integral

$$I_1 = \int_B^A \frac{f(z)}{\sqrt{z - a} + b} dz$$

[1]The reader may note that any line joining the origin and a point on the circle of infinite radius could be a branch cut for the function $f(z) = \sqrt{z}$.

On the line AB, the argument of $(z - a)$ is positive. Therefore, in this path we consider $\sqrt{z-a} = r\exp(j\theta/2)$, where $r = |z - a|^{1/2}$ and θ is the principle value of the argument $(z - a)$. Also, $0 < \theta < 2\pi$ in this case. Now consider the integral

$$I_3 = \int_{A'}^{B'} \frac{f(z)}{\sqrt{z-a} + b}\, dz$$

From Figure A3.2, we see that $z - a$ changes its phase by $-360°$ (the negative sign is due to a clockwise rotation), where z moves from A to A' along the circumference of the small circle. Therefore, at A'. we express $z - a = r\exp(j(\theta - 2\pi))$, so that

$$\sqrt{z-a}\big|_{\text{at } A'} = \sqrt{r}\exp(j\theta/2 - j\pi) = -\sqrt{r}\exp(j\theta/2) = -\sqrt{z-a}\big|_{\text{at } A}$$

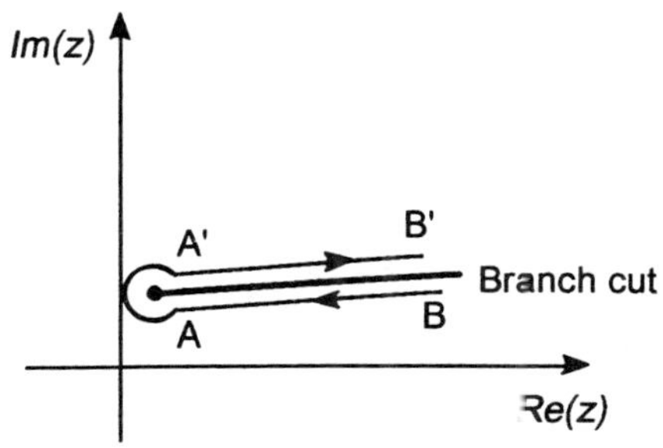

Figure A3.2 An integration path around a branch cut.

In fact the arguments of $\sqrt{z-a}$ on $A'B'$ and AB differ by 180° for all z. Therefore, we obtain

$$I_3 = \int_{A}^{B} \frac{f(z)}{-\sqrt{z-a} + b}\, dz$$

Now consider the integral I_2 on the circular contours and substitute

$$z - a = r\exp(j\theta)$$

$$dz = jr\exp(j\theta)\, d\theta$$

$$I_2 = \int_{\theta_A}^{\theta_A - 2\pi} \frac{f[a + r\exp(j\theta)]}{\sqrt{r}\exp(j\theta/2) + b}\, jr\exp(j\theta)\, d\theta$$

When $r \to 0$, $I_2 \to 0$ if $|b| > 0$. Therefore,

$$I = I_1 + I_2 = \int_B^A \left\{ \frac{f(z)}{\sqrt{z-a}+b} - \frac{f(z)}{b-\sqrt{z-a}} \right\} dz$$

$$= 2\int_B^A \frac{f(z)\sqrt{z-a}}{z-a-b^2}\, dz$$

This integral is a simple line integral suitable for numerical computation.

Practice Problem. Using the method of residue calculus, evaluate the closed contour integral

$$\oint_c \frac{z^{\alpha-1}}{1+z}\, dz \qquad 0 < \alpha < 1$$

where the contour is shown in Figure A3.3. Hence, prove that

$$\int_0^\infty \frac{z^{\alpha-1}}{1+z}\, dz = \frac{\pi}{\sin\pi\alpha} \qquad 0 < \alpha < 1$$

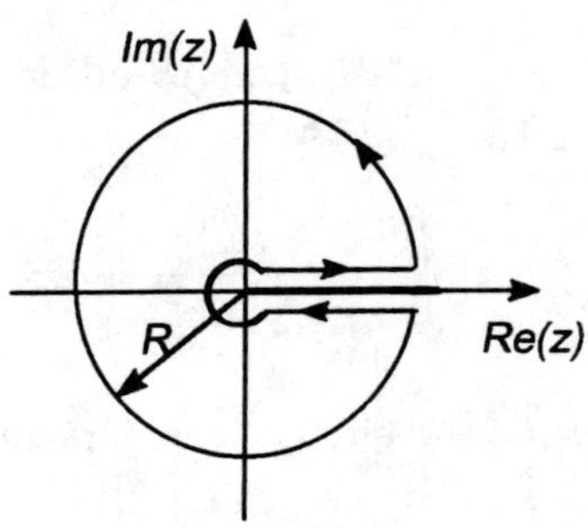

Figure A3.3 An integration contour with $R \to \infty$. The branch cut is along the positive real axis.

List of Symbols

$\mathbf{A}$	Magnetic vector potential
A_z	z-component of the magnetic vector potential
$\tilde{A}_z$	A_z in spectral domain
A_z^{sp}	A_z for the space wave field
A_{zi}^{su}	A_z for the ith surface wave mode
$\mathbf{B}$	Magnetic displacement vector
$\mathbf{B}_t$	Transverse magnetic displacement vector
B_z	z-component of the magnetic displacement vector
B_z^{sp}	B_z for the space wave field
B_{zi}^{su}	B_z for the ith surface wave mode
C	Capacitance per unit length of a transmission line
c_f	Fringing field capacitance
$\mathbf{D}$	Electric displacement vector
D_z	z-component of the electric displacement vector
D_z^{sp}	D_z for the space wave field
D_{zi}^{su}	D_z for the ith surface wave mode
d_i	Width of the ith layer
$\mathbf{E}$	Electric field vector
E_x, E_y, E_z	Electric field components in Cartesian coordinate system
E_ρ, E_ϕ, E_z	Electric field components in cylindrical coordinate system
$\mathbf{E}_t$	Transverse electric field vector
E_y^{inc}	y-component of the incident electric field
$\mathbf{F}$	Electric vector potential
F_z	z-component of the electric vector potential
$\tilde{F}_z$	F_z in spectral domain
F_z^{sp}	F_z for space wave field
F_{zi}^{su}	F_z for the ith surface wave mode

$F\{f(x, y)\}$	Fourier transform of the function $f(x, y)$
$\tilde{f}$	Function $f(x, y)$ in spectral domain
g_{sp}	Edge conductance contributed by the space wave fields
g_{su}	Edge conductance contributed by the surface wave fields
$\mathbf{H}$	Magnetic field vector
$\mathbf{H}_t$	Transverse magnetic field vector
H_x, H_y, H_z	Magnetic field components in Cartesian coordinate system
H_ρ, H_ϕ, H_z	Magnetic field components in a cylindrical coordinate system
$H_n^{(1)}(x), H_n^{(2)}(x)$	nth-order Hankel functions of first and second kind
$H_n^{(1)'}(x), H_n^{(2)'}(x)$	Derivatives of Hankel functions with respect to x
H_x^{inc}	x-component of the incident electric field
h	Substrate thickness
I_p	Probe current
$J_n(x)$	Bessel function of order n
$J_n'(x)$	Derivative of $J_n(x)$ with respect to x
$\mathbf{J}_e$	Electric volume current density
$\mathbf{J}_m$	Magnetic volume current density
$\mathbf{J}_{se}$	Electric surface current density
$\mathbf{J}_{sm}$	Magnetic surface current density
j	$\sqrt{-1}$
k	Wave number
k_0	Wave number in free space
k_i	Wave number in the ith layer
k_x, k_y, k_z	Wave number along x, y, and z
$k_{\rho i}$	Radial wave number for the ith surface wave mode
k_{iz}	Wave number in the z direction for the ith layer
k_{1zi}	Wave number along z for the ith surface wave mode in the dielectric region
L	Inductance per unit length of a transmission line
$\mathbf{M}$	Magnetic current
$\hat{n}$	Unit vector along the normal direction of a surface
P^{sp}	Space wave power
P_i^{su}	Surface wave power for the ith surface wave mode
Q	Electric charge
$\text{res}\{f(z), z_0\}$	Residue of $f(z)$ at $z = z_0$
$\langle s_1, s_2 \rangle$	Mutual reaction between two sources s_1 and s_2
T_0	Transmission coefficient
$T_n(x)$	Tchebysheff polynomial of x
$u(x)$	Unit step function of x
V_0	Source voltage
$Y_n(x)$	Neumann function of order n

$Y'_n(x)$	Derivative of $Y_n(x)$ with respect to x
y^e_{0i}	Characteristic admittance of the transmission line associated with the ith layer for the TE_z mode
y^m_{0i}	Characteristic admittance of the transmission line associated with the ith layer for the TM_z mode
y^m_{in}, y^e_{in}	Input admittance at the source location of the TM_z and TE_z transmission line equivalent circuits
y_n	Edge admittance of a patch antenna
y_r	Radiation admittance
y^{strip}	Equivalent admittance of a conducting strip
Z_0	Characteristic impedance of a transmission line
$Z^{(n)}_{\text{in}}$	Input impedance for the nth mode
Z_{in}	Total input impedance
β	Propagation constant of a transmission line
$\delta(z)$	Dirac-delta function of z
∇_t	Transverse gradient operator $= \nabla - \hat{z}(\partial/\partial z)$
ε_0	Permittivity of free space
ε	Permittivity of a medium
ε_r	Dielectric constant or relative permittivity
ε_i	Permittivity for the ith layer
Φ	Electrostatic potential
γ_{0zi}	Attenuation constant along z of the ith surface wave mode in the free-space region
Γ_i	Reflection coefficient
η	Free-space impedance $= 120\pi$ ohms
λ_0	Wavelength in free space
μ_0	Permeability of free space
μ_i	Permeability of the ith layer
μ_r	Relative permeability
ρ_e, ρ_m	Electric and magnetic volume charge density
ρ_{se}, ρ_{sm}	Electric and magnetic surface charge density
σ_n	Equals 2 for $n = 0$, 1 for $n \neq 0$
ω	Angular frequency

Bibliography

ELECTROMAGNETICS

J. Van Bladel, *Electromagnetic Fields*, McGraw-Hill, New York, 1964.

R. E. Collin, *Field Theory of Guided Waves*, 2nd. ed., IEEE Press, New York, 1990.

R. F. Harrington, *Time-Harmonic Electromagnetic Fields*, McGraw-Hill, New York, 1961.

J. A. Kong, *Electromagnetic Wave Theory*, John Wiley and Sons, New York, 1986.

S. A. Schelkunoff, *Electromagnetic Fields*, Blaisdell Publishing Company, 1963.

J. A. Stratton, *Electromagnetic Theory*, McGraw-Hill, New York, 1941.

SPECTRAL DOMAIN ANALYSIS AND WAVES IN LAYERED MEDIA

K. Araki and T. Itoh, "Hankel Transform Domain Analysis of Open Circular Microstrip Radiating Structures," *IEEE Trans. on Antennas and Propagation*, Vol. AP-29, 1981, pp. 84–89.

A. K. Bhattacharyya, "Characteristics of Space and Surface Waves in a Multilayered Structure," *IEEE Trans. on Antennas and Propagation*, Vol. AP-38, 1990, pp. 1231–1238.

A. K. Bhattacharyya and R. Garg, "Spectral Domain Analysis of Wall Admittances for Circular and Annular Patches and the Effect of Surface Waves," *IEEE Trans. on Antennas and Propagation*, Vol. AP-33, 1985, pp. 1067–1073.

L. M. Brekhovskikh, *Waves in Layered Media*, Academic Press, New York, 1960.

W. C. Chew and J. A. Kong, "Analysis of a circular Microstrip Disk Antenna with a Thick Dielectric Substrate," *IEEE Trans. on Antennas and Propagation*, Vol. AP-29, 1981, pp. 68–76.

R. E. Collin, *Field Theory of Guided Waves*, 2d ed., IEEE Press, New York, 1990, Chapters 4 and 11.

L. B. Felsen and N. Marcuvitz, *Radiation and Scattering of Waves*, Prentice-Hall, Englewood Cliffs, NJ, 1973.

T. Itoh and W. Menzel, "A Full Wave Analysis of Open Microstrip Structures," *IEEE Trans. on Antennas and Propagation*, Vol. AP-29, Jan. 1981, pp. 63–68.

T. Itoh and R. Mittra, "Spectral-Domain Approach for Calculating Dispersion Characteristics of Microstrip Lines," *IEEE Trans. on Microwave Theory and Techniques*, Vol. MTT-21, 1973, pp. 496–498.

D. R. Jackson and A. A. Oliner, "A Leaky-Wave Analysis of High Grain Printed Antenna Configuration," *IEEE Trans. on Antennas and Propagation*, Vol. AP-36, July 1988, pp. 905–910.

P. B. Katchi and N. G. Alexopoulos, "Real Axis Integration of Sommerfeld Integrals with Applications to Printed Circuit Antennas," *J. Math. Phys.*, Vol.-14, 1983, pp. 527–533.

P. B. Katchi and N. G. Alexopoulos, "On the Modeling of Electromagnetically Coupled Microstrip Antennas—The Printed Strip Dipole," *IEEE Trans. on Antennas and Propagation*, Vol. AP-32, 1984, pp. 1179–1186.

J. A. Kong, *Electromagnetic Wave Theory*, John Wiley and Sons, New York, 1986, Chapter 4.

N. Marcuvitz, "On Field Representations in Terms of Leaky Modes or Eigenmodes," *Trans. Inst. Radio Engrs.*, Vol. AP-4, 1956, pp. 192–194.

J. R. Mosig and F. E. Gardiol, "General Integral Equation Formulation for Microstrip Antennas and Scatterers," *Proc. IEE*, Part H, Vol. 132, 1985, pp. 424–432.

D. M. Pozar and S. M. Voda, "A Rigorous Analysis of Microstrip Fed Patch Antenna," *IEEE Trans. on Antennas and Propagation*, Vol. AP-35, 1987, pp. 1343–1350.

I. E. Rana and N. G. Alexopoulos, "Current Distribution and Input Impedance of Printed Dipoles," *IEEE Trans. on Antennas and Propagation*, Vol. AP-29, 1981, pp. 99–105.

J. A. Stratton, *Electromagnetic Theory*, McGraw-Hill, New York, 1941, Chapter 9.

T. Tamir and A. A. Oliner, "Guided Complex Waves Part 1, Fields at an Interface," *Proc. IEE*, Vol. 110, 1963, pp. 310–324.

T. Tamir and A. A. Oliner, "Guided Complex Waves Part 2, Fields at an Interface," *Proc. IEE*, Vol. 110, 1963, pp. 325–334.

J. R. Wait, *Electromagnetic Waves in Stratified Media*, 2d ed., Pergamon Press, New York, 1970.

MICROSTRIP ANTENNA ELEMENTS

N. G. Alexopoulos, P. B. Katehi, and D. B. Rutledge, "Substrate Optimization for Integrated Circuit Antennas," *IEEE Trans. on Microwave Theory and Techniques*, Vol. MTT-31, 1983, pp. 550–557.

M. C. Bailey and M. D. Deshpande, "Integral Equation Formulation of Microstrip Antennas," *IEEE Trans. on Antennas and Propagation*, Vol. AP-30, 1982, pp. 651–656.

I. J. Bhal and P. Bhartia, *Microstrip Antennas*, Artech House, Dedham, MA, 1980.

A. K. Bhattacharyya, "Analysis of Circular Patch Antennas on Electrically Thick Substrate," *Computer Physics Communications*, 1991, pp. 485–495.

A. K. Bhattacharyya, "Characteristics of Circular Patch on Thick Substrate and Superstrate," *IEEE Trans. on Antennas and Propagation*, Vol. AP-39, 1991, pp. 1038–1041.

A. K. Bhattacharyya, "Effects of Ground Plane and Dielectric Truncation on the Efficiency of a Printed Structure," *IEEE Trans. on Antennas and Propagation*, Vol. AP-39, 1991, pp. 303–308.

A. K. Bhattacharyya, "Effects of Ground Plane Truncation on the Impedance of a Patch Antenna," *Proc. IEE*, Part H, Vol. 138, 1991, pp. 560–564.

A. K. Bhattacharyya and R. Garg, "Generalized Transmission Line Model for Microstrip Patches," *IEE Proc.*, Part H, Vol. 132, 1985, pp. 93–98.

A. K. Bhattacharyya and R. Garg, "Effect of Substrate on the Efficiency of an Arbitrarily Shaped Microstrip Patch Antenna," *IEEE Trans. on Antennas and Propagation*, Vol. AP-34, 1986, pp. 1181–1188.

A. K. Bhattacharyya, Y. M. M. Antar, and A. Ittipiboon, "Spectral Domain Analysis of Aperture-Coupled Microstrip Patch Antennas," *Proc. IEE*, Part H, Vol. 139, 1992, pp. 459–464.

A. G. Derneryd, "Linearly Polarized Microstrip Antennas," *IEEE Trans. on Antennas and Propagation*, Vol. AP-24, 1976, pp. 846–851.

S. B. D. Fonseca and A. J. Giarola, "Microstrip Disk Antennas, Part I: Efficiency of Space Wave Launching," *IEEE Trans.*, Vol. AP-32, 1984, pp. 561–567.

K. C. Gupta and A. Benalla, *Microstrip Antenna Design*, Artech House, Norwood, MA, 1988.

IEEE Trans. on Antennas and Propagation, (special issue on microstrip antennas), Vol. AP-29, January 1981.

D. R. Jackson and N. G. Alexopoulos, "Gain Enhancement Method for Printed Circuit Antennas," *IEEE Trans. on Antennas and Propagation*, Vol. AP-33, 1985, pp. 976–987.

J. R. James, P. S. Hall, and C. Wood, *Microstrip Antenna Theory and Design*, Peter Peregrinus, Stevenage, UK, 1981.

A. A. Kishk and L. Shafai, "The Effect of Various Parameters of Circular Microstrip Antennas on Their Radiation Efficiency and the Mode Excitation," *IEEE Trans. on Antennas and Propagation*, Vol. AP-34, 1986, pp. 969–976.

D. M. Pozar, "Considerations for Millimeter Wave Printed Antennas," *IEEE Trans. on Antennas and Propagation*, Vol. AP-31, 1983, pp. 740–747.

D. M. Pozar, "A Reciprocity Method of Analysis Printed Slot and Slot-Coupled Microstrip Antenna," *IEEE Trans. on Antennas and Propagation*, Vol. AP-34, 1986, pp. 1439–1446.

MICROSTRIP ARRAYS

A. K. Bhattacharyya and L. Shafai, "Effect of Mutual Coupling on the Radiation Pattern of Microstrip Phased Array Antenna," *IEEE AP-S Int. Symp. Dig.*, 1986, pp. 891–893.

P. S. Hall, "Microstrip Linear Array with Polarization Control," *Proc. IEE*, Part H, Vol. 130, 1983, pp. 215–224.

D. R. Jackson and P. Manghnani, "Analysis and Design of Linear Array of Electromagnetically Coupled Microstrip Patches," *IEEE Trans. on Antennas and Propagation*, Vol. AP-38, 1990, pp. 754–759.

P. B. Katehi, "Mutual Coupling Between Microstrip Dipoles in Multilayered Arrays," *IEEE Trans. on Antennas and Propagation*, Vol. AP-37, 1989, pp. 275–280.

D. M. Pozar, "Infinite Phased Array of Aperture-Coupled Microstrip Patches," *IEEE Trans. on Antennas and Propagation*, Vol. AP-37, 1989, pp. 418–425.

D. M. Pozar and D. H. Schaubert, "Scan Blindness in Infinite Phased Arrays of Printed Dipoles," *IEEE Trans. on Antennas and Propagation*, Vol. AP-32, 1984, pp. 602–610.

D. M. Pozar and D. H. Schaubert, "Analysis of an Infinite Array of Rectangular Microstrip Patches with Idealized Probe Feeds," *IEEE Trans. on Antennas and Propagation*, Vol. AP-32, 1984, pp. 1101–1107.

C. Terret, S. Assailly, K. Mahdjoubi, and M. Edimo, "Mutual Coupling Between Stacked Square Microstrip Antennas fed on their Diagonal," *IEEE Trans. on Antennas and Propagation*, Vol. AP-39, 1991, pp. 1049–1051.

PLANAR TRANSMISSION LINES

Y. M. M. Antar, A. K. Bhattacharyya, and A. Ittipiboon, "Microstripline-Slotline Transition Analysis Using the Spectral Domain Technique," *IEEE Trans. on Microwave Theory and Techniques*, Vol. MTT-40, 1992, pp. 515–523.

H. G. Bergandt and R. Pregla, "Calculation of Even and Odd-mode Capacitance Parameters for Coupled Microstrips," AEU, Vol. 26, 1972, pp. 153–158.

S. B. Cohn, "Slotline on a Dielectric Substrate," *IEEE Trans. on Microwave Theory and Techniques*, Vol. MTT-17, 1969, pp. 768–778.

S. B. Cohn, "Slotline Field Components," *IEEE Trans. on Microwave Theory and Techniques*, Vol. MTT-20, 1972, pp. 172–174.

G. M. L. Gladwell and S. Coen, "A Chebyshev Approximation of Microstrip Transmission Lines," *IEEE Trans. on Microwave Theory and Techniques*, Vol. MTT-23, 1975, pp. 865–876.

R. Garg, "Effect of Tolerances on Microstripline and Slotline Performances," *IEEE Trans. on Microwave Theory and Techniques*, Vol. MTT-26, 1978, pp. 16–19.

R. Garg and K. C. Gupta, "Expressions for Wavelength and Impedance of Slotline," *IEEE Trans. on Microwave Theory and Techniques*, Vol. MTT-24, 1976, p. 532.

W. J. Getsinger, "Dispersion of Parallel-Coupled Microstrip," *IEEE Trans. on Microwave Theory and Techniques*, Vol. MTT-21, 1973, pp. 144–145.

W. J. Getsinger, "Microstrip Dispersion Model," *IEEE Trans. on Microwave Theory and Techniques*, Vol. MTT-21, 1973, pp. 34–39.

K. C. Gupta, R. Garg, and I. J. Bahl, *Microstrip Lines and Slot Lines*, Artech House, Dedham, MA, 1979.

J. S. Hornsby and A. Gopinath, "Fourier Analysis of a Dielectric-loaded Waveguide with a Microstrip," *Electronics Letter*, Vol. 5, 1969, pp. 265–267.

T. Itoh and R. Mittra, "Dispersion Characteristics of Slotlines," *Electronics Letter*, Vol. 7, 1971, pp. 364–365.

T. Itoh and R. Mittra, "Spectral-Domain Approach for Calculating Dispersion Characteristics of Microstrip Lines," *IEEE Trans. on Microwave Theory and Techniques*, Vol. MTT-21, 1973, pp. 496–498.

J. B. Knorr, "Slot-Line Transitions," *IEEE Trans. on Microwave Theory and Techniques*, Vol. MTT-22, 1974, pp. 548–554.

J. B. Knorr and K. D. Kuchler, "Analysis of Coupled Slots and Coplanar Strips on Dielectric Substrates," *IEEE Trans. on Microwave Theory and Techniques*, Vol. MTT-23, 1975, pp. 541–548.

J. B. Knorr and A. Tufekcioglu, "Spectral-Domain Calculation of Microstrip Characteristics Impedance," *IEEE Trans. on Microwave Theory and Techniques*, Vol. MTT-23, 1975, pp. 725–728.

G. Kowalski and R. Pregla, "Dispersion Characteristics of Single and Coupled Microstrips," AEU, Vol. 26, 1972, pp. 276–280.

R. Mittra and T. Itoh, "A New Technique for the Analysis of the Dispersion Characteristics of Microstrip Lines," *IEEE Trans. on Microwave Theory and Techniques*, Vol. MTT-19, 1971, pp. 47–56.

P. Silvester, "TEM Properties of Microstrip Transmission Lines," *Proc. IEE*, Vol. 115, 1968, pp. 42–49.

E. Yamashita, "Variational Method for the Analysis of Microstrip-like Transmission Lines," *IEEE Trans. on Microwave Theory and Techniques*, Vol. MTT-16, 1968, pp. 529–535.

E. Yamashita and R. Mittra, "Variational Method for the Analysis of Microstrip Lines," *IEEE Trans. on Microwave Theory and Techniques*, Vol. MTT-16, 1968, pp. 251–256.

MATHEMATICS

E. Kreyszig, *Advanced Engineering Mathematics*, 6th ed., John Wiley, New York, 1988.

J. Mathews and R. L. Walker, *Mathematical Methods of Physics*, 2nd ed., W. A. Benjamin, Menlo Park, CA, 1979.

M. R. Spiegel, *Vector Analysis*, Schaum's Outline Series, McGraw-Hill, New York, 1974.

M. R. Spiegel, *Complex Variables*, Schaum's Outline Series, McGraw-Hill, Singapore, 1981.

A. G. Sveshnikov and A. N. Tikhonov, *The Theory of Functions of a Complex Variable*, Mir Publishers, Moscow, 1978.

Index

The Artech House Antenna Library

Helmut E. Schrank, *Series Editor*

Advanced Technology in Satellite Communication Antennas: Electrical and Mechanical Design, Takashi Kitsuregawa

Analysis, Design, and Measurement of Small and Low-Profile Antennas, K. Hirasawa

Analysis of Wire Antennas and Scatterers: Software and User's Manual, A. R. Djordjevic, M. B. Bazdar, G. M. Bazdar, G. M. Vitosevic, T. K. Sarkar, and R. F. Harrington

Analysis Methods for Electromagnetic Wave Problems, E. Yamashita, editor

Antenna Measurement Techniques, Gary E. Evans

Antenna-Based Signal Processing Techniques for Radar Systems, Alfonso Farina

CAD for Linear and Planar Antenna Arrays of Various Radiating Elements: Software and User's Manual, Aleksandar Nesic and Miodrag Mikavica

Computational Electromagnetics, Konada Umashankar

Designer Notes for Microwave Antennas, Richard C. Johnson

Fixed and Mobile Terminal Antennas, A. Kumar

Generalized Multipole Technique for Computational Electromagnetics, Cristian Hafner

Integral Equation Methods for Electromagnetics, N. Morita, N. Kumagai, and J. Mautz

Four-Armed Spiral Antennas, Robert G. Corzine and Joseph A. Mosko

Interference Suppression Techniques for Microwave Antennas and Transmitters, Ernest Freeman

Introduction to Electromagnetic Wave Propagation, Paul Rohan

Introduction to the Uniform Geometrical Theory of Diffraction, D. A. McNamara

Microwave Cavity Antennas, A. Kumar and H. D. Hristov

Millimeter-Wave Microstrip and Printed Circuit Antennas, Prakash Bhartia

Mobile Antenna Systems, K. Fujimoto and J. R. James

Modern Methods of Reflector Antenna Analysis and Design, Craig Scott

Moment Methods in Antennas and Scattering, Robert C. Hansen, editor

Monopole Elements on Circular Ground Planes, M. M. Weiner *et al.*

Near-Field Antenna Measurements, D. Slater

Phased Array Antenna Handbook, Robert J. Mailloux

Polariztion in Electromagnetic Systems, Warren Stutzman

Practical Phased-Array Antenna Systems, Eli Brookner *et al.*

Practical Simulation of Radar Antennas and Radomes, Herbert L. Hirsch and Douglas C. Grove

Reflector and Lens Antenna Analysis and Design Using Personal Computers, Carlyle J. Sletten, editor

Shipboard Antennas, Second Edition, Preston Law

Small-Aperture Radio Direction-Finding, Herndon Jenkins

Spectral Domain Method in Electromagnetics, Craig Scott

Two-Dimensional Multiple Multipole Analysis Software and User's Manual, Christian Hafner

Waveguide Components for Antenna Feed Systems: Theory and CAD, J. Uher, J. Bornemann, and Uwe Rosenberg

For further information on these and other Artech House titles, contact:

Artech House
685 Canton Street
Norwood, MA 01602
617-769-9750
Fax: 617-762-9230
Telex: 951-659
email: artech@world.std.com

Artech House
6 Buckingham Gate
London SW1E6JP England
+44 (0) 71-973-8077
Fax: +44 (0) 71-630-0166
Telex: 951-659